Collins *gem*

Cocktails

D0243458

For Sarah, always my favourite cocktail companion

This revised edition published 2004 by Collins
an imprint of HarperCollins*Publishers*
75–85 Fulham Palace Road, London W6 8JB

First published 1999

The right of Jeremy Harwood to be identified as the author
of this work has been asserted by him in accordance with
the Copyright, Designs and Patents Act 1988.

Collins Gem® is a registered trademark of
HarperCollins*Publishers* Limited

© Essential Works 1999, 2004

A CIP catalogue record for this book is available from the
British Library.

ISBN 0 00 719078 6

Colour reproduction by Digital Imaging
Printed in Italy by Amadeus S.p.A.

The authors and publishers have made every reasonable
effort to contact all copyright holders. Any errors that may
have occurred are inadvertent and anyone who for any
reason has not been contacted is invited to write to the
publishers so that a full acknowledgement may be made
in subsequent editions of this work.

CONTENTS

INTRODUCTION   **8**

ENTER THE COCKTAIL   10

WHAT MAKES A GOOD COCKTAIL?   13

WHAT YOU WILL NEED   16

STOCKING UP THE BAR   17

WHICH SPIRITS?   18

WHICH LIQUEURS?   21

WHICH MIXERS?   24

WHICH FIXINGS?   25

WHY ICE?   27

WHICH KIT?   28

WHICH GLASSES?   31

SHAKE, STIR OR POUR?   33

RUM-BASED   **36**

PLANTERS'S PUNCH   39

BRASS MONKEY   40

ZOMBIE   41

CUBA LIBRE   43

CORKSCREW   44

PIÑA COLADA   44

BACARDI   45

BEE'S KNEES   46

CLASSIC DAIQUIRI   47

BANANA DAIQUIRI   49

BETWEEN THE SHEETS   50

TOM AND JERRY   50

MOJITO   52

ACAPULCO   53

MARY PICKFORD   54

MAI TAI   56

LONG ISLAND TEA   57

BRANDY-BASED   **58**

B&B   61

BRANDY ALEXANDER   62

CHARLESTON   63

AU REVOIR   64

EGGNOG   66

BRANDY CLASSIC   67

SIDECAR   68

BRANDY SOUR   69

HORSE'S NECK   71

STINGER   71

GRENADIER   72

COFFEE   73

HARVARD   74

BOSOM CARESSER   75

BLACKSMITH COCKTAIL   76

BOMBAY COCKTAIL   77

GIN-BASED   **78**

PINK LADY   81

WHITE LADY   83

GIN SLING	84
GIN AND SIN	85
GIN SOUR	86
GIN SMASH	87
GIN FIZZ	88
CARDINALE	89
NEGRONI	90
CARUSO	93
SNOWBALL	94
BLUE DEVIL	94
BRONX	96
GIN ALEXANDER	97
TOM COLLINS	97
SINGAPORE SLING	100
MAIDEN'S PRAYER	102
FRENCH 75	102
GIMLET	103
GREEN DEVIL	104
GIBSON	105
PINK GIN	107
VODKA-BASED	**108**
BLOODY MARY	111
BLUE LAGOON	113
MOSCOW MULE	114
COSMOPOLITAN	115
ORANGE COSMO	116
THE BLACK MARBLE	116
SALTY DOG	117

HARVEY WALLBANGER	118
VODKA MARTINI	120
LONG VODKA	121
SOVIET	121
SCREWDRIVER	122
WHITE RUSSIAN	123
THE CLASSIC MARTINI	**124**
DRY MARTINI	126
MEDIUM MARTINI	126
SWEET MARTINI	127
APPLE MARTINI	127
WHISKY-BASED	**128**
OLD-FASHIONED	131
ROB ROY	132
THREE RIVERS	133
WHISKY SOUR	134
SLOW COMFORTABLE SCREW	136
WHISKY HIGHBALL	137
MANHATTAN	138
MINT JULEP	140
NEW ORLEANS	141
ALGONQUIN	142
EDINBURGH	143
WALDORF	144
BRAINSTORM	145
RUSTY NAIL	146
REBEL CHARGE	147

CHAMPAGNE-BASED 148
CHAMPAGNE COCKTAIL 151
BELLINI 153
KIR ROYALE 154
BUCK'S FIZZ 155
BLACK VELVET 156
ROSSINI 158
DEATH IN THE AFTERNOON 159
TYPHOON 160
ARISE MY LOVE 161
AMBROSIA 163

SHOOTERS, SLAMMERS AND TEQUILA 164
POUSSÉ CAFÉ 166
TEQUILA SLAMMER 168
MARGARITA 170
TAMARIND MARGARITA 171
TEQUILA SUNRISE 172

NON-ALCOHOLIC 174
VIRGIN MARY 177
SHIRLEY TEMPLE 178
ST CLEMENTS 180
GINGER FRUIT PUNCH 181

HANGOVER CURES 182
PRAIRIE OYSTER 184
BULLSHOT 186
SUNBURST 186

INDEX 188

BOTTOMS UP!

No one knows exactly where, why, when, or by whom the first cocktails were mixed and shaken, though a likely candidate for the honour is Betsy Flanagan, an enterprising Irish-American innkeeper, who, it is said, decorated the bottles from which she decanted her concoctions with gaily coloured tail feathers from fighting cocks. The story goes that a French customer of hers was so taken with this unusual presentation that he raised his glass to Betsy with the toast 'Vive le cocktail'!

Whether this tale is true or not remains a matter of conjecture. Some cocktail buffs hold that the name really comes from cock-ale, a mixture of spirits administered to fighting cocks in 18th-century England to get their courage up before their bouts, while others argue that the name was derived from a mixed drink called a coquetel, which originated in the Bordeaux region of France. Apparently, this was a favourite tipple among the French officers serving alongside the rebellious colonists during the American War of Independence, in what were to become the southern states of the U.S.A. What cannot be disputed, however, is that cocktails started

their real climb to fame in 20s America, when they became the preferred illegal beverage for the hundreds of thousands of drinkers confronting the vigours and rigours of Prohibition. A total ban on the 'manufacture, sale and transport of alcoholic drinks' held the nation in its grip from 1919, when the 18th Amendment to the Constitution was passed into law over President Woodrow Wilson's veto, until 1933, when it once again became legal to drink alcohol in the U.S.A.

ENTER THE COCKTAIL

In 1928, President-elect Herbert Hoover may still have been referring to Prohibition as 'a great social and economic experiment, noble in motive and far-reaching in purpose', but, by this time, the whole affair was already proving itself to be an abject fiasco. Despite all the efforts of the police, Customs officers and revenue officials, bootlegging had become big business, with organized crime heavily involved: in towns and cities throughout the U.S.A., illicit speakeasies had taken over from traditional bars and saloons (the name 'speakeasy' probably comes from the security governing admission to the premises).

Once the drinkers were inside, the quality of the liquor, in the main, was nothing to write home about. To satisfy a nation that seemed to be growing thirstier by the minute – and, at the same time, maximise their own profits – the bootleggers were being driven to setting up hidden stills of their own to supplement the shiploads and truckloads of booze that they were smuggling into the country from abroad. Soon, they were producing gallon after gallon of their own disreputable versions of whisky, gin and rum: 'bathtub' gin was one favourite, with 'coffin varnish', 'rotgut' and 'moonshine' being among the others. The one quality they all had in common was that, as raw spirits, they tasted truly dreadful. At their worst, they actually could be downright poisonous.

With bars across the country searching for ways of making what they were selling palatable, cocktails were quick to come into their own. Soon, they became all the rage and expert bartenders (or mixologists, as they became known in the U.S.A.) became living legends on the strength of the cocktail recipes they originated. From the U.S.A., the fashion spread to Europe. London, Paris, Nice, Venice – to keep up with the times, any place with the slightest claim to being in the swing of things had to follow the trend. Just as the bar of the St Regis Hotel became, for a time, the place for cocktails in New York, in London there was the American Bar at the Savoy and the Rivoli Bar at the Ritz. Paris and Venice each had their own Harry's Bar – the former was one of Ernest Hemingway's favourite watering-holes – while out East, in the heart of colonial Singapore, there was the Long Bar at Raffles Hotel, birthplace of the immortal Singapore Sling.

Where cocktail bars led, the soon-to-be-as-popular notion of the cocktail party followed. This was espoused equally eagerly by the jazz-loving, cigarette-holder-flourishing, Charleston-dancing, cocktail-sinking Bright Young Things of the period. For it was the young that took to the cocktail habit the most keenly. The what-was-good-enough-for-father brigade might still stick to sherry as the traditional aperitif before dinner, but why serve boring old sherry when there was a host of wonderfully exotic alternatives to shake, mix and pour? After all, this was the generation that re-dubbed the Boer War the 'Bore War' and turned 'England's green and pleasant land' into 'England's green unpleasant land'. For Noel Coward's Amanda and Elyot, there was nothing to beat sipping a champagne cocktail as the sun slowly set at fashionable Deauville, while, back in London, the inimitable Jeeves would be mixing the driest of Dry Martinis – history, unfortunately, does not record the exact recipe – as Bertie Wooster's indispensable pre-prandial tincture before the latter departed to meet his fellow 'eggs, beans and crumpets' at the immortal Drones Club.

WHAT MAKES A GOOD COCKTAIL?

Today, there are literally thousands of different cocktails, enough to suit every age, taste and occasion. They range from elegant classics, such as the Dry Martini, Manhattan and Sidecar, to such colourful and exotic creations as the Harvey Wallbanger, Piña Colada and Tequila Sunrise, to the contemporary Cosmopolitan and Mojito. However, there is simply no right or wrong way of making these and all the other different varieties. What there is, as cocktail fans will undoubtedly tell you, is a mystique that has sprung up over the years about the virtues of those closely-guarded individual touches that individual bartenders have injected into the recipes. For the key to a truly memorable cocktail is undoubtedly

Cocktail shaker

individuality. This is why the search for that single 'greatest Dry Martini in the world' is somewhat like the epic quest on which King Arthur and the Knights of the Round Table embarked in search of the Holy Grail. It will probably take just as long – and end just as unsuccessfully. Each and every bartender has his or her own view and version: one celebrated American bartender holds, for instance, that the secret lies in careful preparation of a Martini glass. What to do is to pour no more than a quarter capful of vermouth into the glass, swill it around and empty it out before the glass is chilled.

In fact, there is little or no magic involved when it comes to the basics of mixing a good cocktail. It is much more of a science than an art – and the first principles are easy to master. With a smattering of basic knowledge and a dash of commonsense, it is by no means difficult to prepare cocktails that will tickle the fancy of even the most discriminating or jaded of palates. For instance, whether you shake, stir or pour is determined by what the ingredients are, as they determine the choice of method. If fruit juice, egg white, egg yolk, cream, milk or other cloudy or opaque ingredients are involved, the use of a cocktail shaker is called for, according to the experts. If all the ingredients are clear, then the way to mix them is to stir them together in a bar glass.

Remember, too, that it can be easy to get confused by quantities – especially since, depending on nationality or location, different bartenders can use different systems of measurement. This is why it is always best to work in proportions, if you can. Finally, as you gain in confidence, you can be more inventive. The recipes given later in this book are accepted cocktail recipes, but you could experiment with lots of little variations within them. A touch of Grenadine or sugar syrup can supply a touch of extra sweetness, while a dash of Angostura bitters, lemon juice or lime juice will give you an added hint of bitterness or sourness respectively. If you are lucky, you will discover that, like all the great bartenders, you have the knack of hitting on exactly the right tricks to create something truly individual.

What it really comes down to is that, as far as cocktail-making is concerned, there is only one golden rule. A good cocktail looks attractive, smells great – and tastes wonderful.

WHAT YOU WILL NEED

Firstly, you will need an assortment of spirits, liqueurs and mixers, along with the fixings that make all the difference to a cocktail's taste and appearance. As far as equipment is concerned, the obvious essential is a cocktail shaker – many professional bartenders today prefer the two-piece Boston version, but the traditional three-piece shaker is easier to handle – and a range of glasses. Cocktails taste better when served ice-cold, so a plentiful supply of ice on tap is also a must.

STOCKING UP THE BAR

Visit any proper cocktail bar, and you may well find yourself overwhelmed by the sheer number of bottles you will see lined up before you, neatly arranged on the bar shelves. Do not be put off by this, especially as far as spirits and liqueurs are concerned. Often, different cocktails have the same basic ingredients – it is the mixes or quantities that differ. This means that, when you are getting started, a relatively small stock will still suffice to enable you to produce an attractive variety. It is a false economy, however, to stint on the quality of what you buy – premium brands undoubtedly produce the best results.

In addition to spirits, you will need a selection of liqueurs, plus a range of mixers and juices ranging from dry and sweet vermouth to fruit juices – tomato juice, for instance, is indispensable if a Bloody Mary is on your list of cocktail priorities. So, too, is soda water, a staple when it comes to making long, thirst-quenching drinks. If gin is your tipple of choice, there is nothing to beat relaxing after a hard, hot summer's day at work with a swiftly-mixed gin fizz or two as a mood-mellowing sundowner.

JANNEAU
VSOP

WHICH SPIRITS?

Basically, you should stock your bar or drinks cabinet with the following:

- **Gin** We prefer Dry London, but each to their own.
- **Vodka** If you can find it, go for the genuine Russian article, although Polish vodka has an equally good 'kick' to it.
- **Rum** You will need both white and dark rum.
- **Scotch whisky** Whatever you do, stick to blends, rather than turning to malts. Any true whisky lover would regard use of the latter as sacrilege.
- **Bourbon** Because many cocktails originated in the U.S.A., this, rather than Scotch, is often the more authentic spirit, though, frankly, unless you are mixing cocktails for a crowd of pedants, Scotch is usually a perfectly acceptable substitute. If, on the other hand, you are a dedicated whisky fan, you could consider ringing the changes still further with Irish whiskey or Canadian Rye.
- **Tequila** At some stage in your cocktail-making career, you're bound to be asked to shake a Margarita or mix a Tequila Sunrise, for which this punchy Mexican spirit is the essential base. If you're brave, you can go for a touch of real old-fashioned authenticity with a few Tequila Slammers, with the salt, slice of lime and a suitable chaser to follow up.

- **Brandy** A reasonable French cognac is the priority here, though, as you get more ambitious, it is a good idea to broaden the selection with a bottle of armagnac and a selection of eaux-de-vie. Calvados is the best-known example of the latter – it is a French fruit brandy (the American equivalent is applejack), distilled from cider. It is worth buying a bottle of this, plus some apricot and cherry brandy as well.

Other essentials include bottles of dry and sweet vermouth. Basically, vermouth is a white wine appetiser, flavoured with many different herbs, roots, berries, flowers and seeds. French vermouth, which is light gold in colour with a nutty flavour, is usually drier than Italian vermouth, which is red, richer and more syrupy. Both varieties are perishable, so they should always be kept in a refrigerator after they have been opened. Otherwise, in time they will lose their freshness. A bottle of Campari is also a good idea – a Campari and soda is an excellent thirst-quencher, when you want something that little bit different, but are not in the cocktail mood – along with a bottle of Pernod or ouzo for the same reasons. Some cocktails call for sherry – the safest bet here is a bottle of amontillado (medium sherry) plus a bottle of fino (dry). You should also invest in a bottle of port – vintage-character port is probably best – and, if you can afford it, real champagne.

Though there are many varieties of sparkling wine – the Californians make a particularly palatable brew – only the French product is legally entitled to be called champagne and, to qualify, it has to come from the champagne region around Rheims, be produced from specified grapes and be allowed to ferment and mature in the traditional way. For cocktails, a dry or dryish champagne – look for the word 'brut' on the bottle – is best. 'Sec', which you will see marked on some labels, is sweeter. NV means non-vintage.

WHICH LIQUEURS?

When it comes to liqueurs, you could be spoilt for choice. The temptation is always to go for too many, rather than too few, the worry being that, without a broad selection to hand, you will find yourself short of a crucial ingredient when it comes to mixing a new cocktail. There is no need to worry, as the selection here will cover most eventualities. While you are about it, remember to raise a glass to toast the saintly bands of medieval monks, who gave the world such alcoholic delights as Benedictine and Green and Yellow Chartreuse.

- **Amaretto** A sweet almond-based Italian liqueur.
- **Anisette** An aniseed-flavoured liqueur.
- **Bailey's** Irish Cream
- **Benedictine** As its name suggests, this French liqueur was invented by a band of Benedictine monks, based at the abbey of Fécamp in Normandy. Golden in colour and brandy-based, it is flavoured with a mix of herbs and honey, although the exact constituents of the mix are still a closely-guarded secret.
- **Chartreuse** Another monastic invention – this time, by the French Carthusians. There are two varieties of this brandy-based liqueur – green and yellow – the former containing more alcohol than the latter. Other ingredients include honey, herbs and spices, with, for the yellow version, the addition of oranges and myrtle.
- **Cointreau** A sweet, sticky orange-flavoured French liqueur.
- **Crème de Cacao** A sweet liqueur with a cocoa-vanilla taste to it.

- **Crème de Cassis** A brandy-based blackcurrant liqueur.
- **Crème de Menthe** A sticky, sweet mint liqueur. Ideally, you will need the white variety as well as the classic green version.
- **Curaçao** Originally from the West Indian island of the same name, this liqueur has a distinct coffee flavour. There are three colours – orange, blue and white. Ideally, you need all three.
- **Drambuie** As you might expect, Scotland's contribution is malt whisky-based, with additional herbs, spices and heather honey.
- **Galliano** An Italian herb liqueur with a characteristic golden colour, flavoured with liquorice and aniseed.
- **Grand Marnier** A French version of Curaçao, which is flavoured with orange peel.
- **Maraschino** A sweet cherry liqueur.

- **Southern Comfort** An American liqueur with a Bourbon base, flavoured with fruit.
- **Tia Maria** A Caribbean liqueur made from rum, coffee extract and spices.
- **Triple Sec** A clear orange liqueur.

WHICH MIXERS?

Again, there is a wide variety to choose from, including orange juice, tomato juice – essential for a Bloody Mary and several other classics – soda water, lemonade and Grenadine. The last is a red non-alcoholic syrup, made from pomegranates. Tonic water, too, is a good mixer, especially suited to gin and vodka, while ginger ale goes well with whisky, Bourbon and gin. You will also need some coconut milk – you can buy this tinned – if you intend to go in for exotic tropical cocktails. The basic requirements are:

- **Club soda**
- **Colas**
- **Ginger ale**
- **Grenadine**
- **Juices** Orange, grapefruit, lemon and lime fruit juices are a must, as are tomato, pineapple and cranberry juices.
- **Mineral water**
- **Tonic water**

WHICH FIXINGS?

By fixings, I mean the bitters, sauces, herbs, spices, garnishes and other trimmings – the little touches that can make all the difference, in cocktail terms, to the end result. The most widely used form of bitters are Angostura bitters, which come from the Caribbean. Unless a recipe tells you otherwise, you should go sparingly with them, using only a few drops at a time. The same advice applies to Tabasco and Worcestershire sauce – the thing to do is to sample as you go along, as you can always add more if the cocktail needs it – and, generally, to all other herbs and spices. In addition, you will need cocktail onions, olives (green and black), lemons, limes, oranges, strawberries, celery stalks, Maraschino cherries, cucumber, pineapple, banana, grated nutmeg, and some mint leaves if you are making

Juleps. Lemon is probably the most frequently used garnish, either in thin slices, in wedges or as a twist which can be dropped into the glass. Crush the peel slightly first as this will help to impart its flavour. Keep supplies of salt, pepper, granulated and caster sugar to hand, as well as cream and extras like papaya juice and passion fruit juice, if tropical-style cocktails are on the agenda. You may also need to prepare some sugar syrup and possibly some sour syrup, which is a mixture of lime and lemon juice. Making sugar syrup is simplicity itself. All you have to do is to heat a mixture of caster sugar and water in a pan – use around half as much sugar as water – stirring all the while with a wooden spoon until the sugar totally dissolves. Bring the mixture to the boil for a few minutes, skim off any scum that appears, let the syrup cool and then pour it into clean, dry airtight bottles. Keep these in a fridge, but for no more than a month.

Finally, don't forget to stock up with swizzle sticks, coasters, cocktail sticks, straws, and, if you really feel the need or are having an 80s party, decorative parasols! Remember, though, that top professional bartenders never, ever, overdress their cocktails – and this goes just as much for fruit garnishes. The aim always should be to enhance the look of the cocktail, not to overwhelm it.

WHY ICE?

Ask any bartender whether he or she could do without ice and they will tell you that a liberal supply is not an option, but an essential. No cocktail can do without it, even if the recipe calls for the ice to be strained off before serving.

The ice you will need takes three forms – ice cubes, crushed ice and cracked ice. As a rule of thumb, most highballs, old-fashioneds and on-the-rocks-drinks call for ice cubes. Cracked ice is best for drinks that require stirring and shaking, while crushed ice is ideal for tall drinks, frappés and cocktails sipped through straws.

In all events, glasses should always be chilled: the golden rule is to 'chill before you fill'. Either you can put the glasses in a refrigerator for a couple of hours in advance of their use, or you can fill them with cracked or crushed ice before mixing the drink. When the drink is ready, empty the glass, shake out all the melted ice and pour in the drink.

WHICH KIT?

To complete your bar, you need a few essential items of bartender's equipment, although if you are not that committed you can substitute ordinary kitchen tools and utensils for many of the genuine articles. If, on the other hand, cocktails are going to feature prominently on your drinking agenda, it is well worth investing in the extra accessories that will not only make your bartending tasks easier, but also enhance your reputation as an on-the-ball host. Of course, the one thing you will need, whatever you decide, is a cocktail shaker, as there really is no practical alternative you can substitute. Also quite a bit of the fun, from the point of view of the drinker, comes from watching a seasoned mixologist shaking away to get the required blend and then pouring the final result. The equipment given here is listed in descending order of importance.

- **Cocktail shaker** Basically, there are two types. The Boston variety, as noted earlier, is the one that many professionals prefer. It consists of two cup-shaped containers, one of glass and one of metal, that fit over each other. The classic shaker, with three rather than two elements, is considered easier for beginners to use, however. The top, secured by a separate tightly-fitting cap and containing a built-in strainer, fits over the bucket-shaped base, which holds the ice and liquid.

Cocktail whisk

Strainer

- **Blender** A goblet blender is the best buy. However, do not be tempted to try to crush ice in it, as this will blunt the blade. Far better to crush ice by hand, wrapping the cubes that need to be crushed in a clean tea towel and crushing them with a wooden kitchen hammer.

- **Measures** It is a good idea to have two measuring jiggers, one twice the capacity of the other. The vital thing is to ensure that, whatever you choose, both the jiggers are calibrated using the same system of measurement – some come marked in gills, some in fluid ounces and others in millilitres.

As long as you remember the golden rule of sticking to proportions, rather than exact quantities, you will avoid confusion.

- **Mixing glass** This is a glass jug without a handle, but with a lip to make pouring easier. The best types come complete with a special type of strainer – this is called a Hawthorn strainer – that fits neatly into the top. You use the same strainer again when decanting the cocktail.
- **Barspoon** This is a spiralled long-handled spoon that reaches right down to the bottom of the mixing glass.

Other useful items include an ice bucket and ice tongs, a corkscrew and bottle opener, a muddler – you use this to crush sugar and mint leaves – a lemon

1　　2　　3　　4

squeezer, bar and canelle knives and a chopping board. The bar knife is used for preparing garnishes and the canelle knife removes the zest from the garnish where necessary. Don't forget clean bar towels and, above all, a couple of damp wiping cloths.

WHICH GLASSES?

Glasses are almost as important a part of the cocktail experience as are cocktails themselves. Indeed, according to the purists, certain cocktails simply have to be served in a specific type of glass: a Martini simply cannot be classed as Martini unless served in a proper Martini glass, for instance. This is taking pedantry to extremes, though it is true to say that certain cocktails are more suited to some glasses than to others. The rule of thumb is that the stronger the drink is, the smaller the glass.

What is important is that any glass you use should be visually appealing and sparkling clean. To check for cleanliness, hold the suspect glass up to the light and, if necessary, repolish it with a dry, fine linen cloth. The best glasses are thin-lipped, transparent, sound off in high registers when 'pinged' with the fingers, and, of course, show off their contents to best effect. Although you could get away with fewer varieties – just the cocktail, highball, lowball glass and the champagne flute – you probably will find a use for all of the following at some time or other:

- **Martini glasses (1)**
- **Champagne saucers (2)**
- **Cocktail glasses (3)**
- **Champagne flutes (4)**
- **Highball/Collins glasses (5, 6)**
- **Lowball/Old-Fashioned glasses (7, 8)**
- **Shot glasses (9)** These miniature glasses are specially suited to shooters, a new breed of cocktail invented, it is claimed, in Canada in the 90s. The spirits and liqueurs that make up the drink are layered, each one being poured into the glass over the back of a spoon. The idea is to swallow the shooter in a single gulp, giving a new twist to the traditional drinker's phrase 'down in one'.

In certain instances, it is traditional for cocktail glasses to be frosted. To frost whole glasses, store them in a refrigerator for long enough to give each glass a white, frosted, ice-cold look and feel. If the rim needs to be sugar-frosted, moisten it with a slice of lime or lemon and dip into caster sugar. Then put the glass back in the refrigerator until it is needed.

For Margaritas, rub the rim of the glass with a lime and substitute coarse salt for the sugar.

SHAKE, STIR OR POUR?

As noted previously, it is the ingredients that determine which method you use to create any specific cocktail. If these include fruit juice, egg white, egg yolk, cream, milk or any cloudy ingredient, then the answer is to shake them together in a cocktail shaker. Don't forget to put ice cubes into the shaker before you add the mix and start to shake: as well as cooling and diluting the mixture slightly, the cubes will act as a beater.

Basically, you should start by filling the shaker half-full with ice cubes, and then add the spirits, mixers and any flavourings. Having put the lid on the shaker, hold the shaker firmly, keeping the lid in place with one hand, and shake vigorously. About ten seconds should be enough for straightforward cocktails, but, if sugar, eggs or syrups are involved, treble the time. The sign to watch out for is the outside of the shaker becoming chill to the touch. Once the cocktail is ready, remove the lid and strain into the glasses.

Cocktails featuring clear ingredients should be stirred in a bar glass. Half-fill the glass with ice before adding the ingredients. You should note that it takes at least ten seconds of stirring to mix a cocktail properly, unless fizzy mixers are involved. In that case, their bubbling will do much of the stirring for you and only two stirs are needed to finish off the job. Pouring is the method of choice when ingredients of different weights are involved and the aim is to produce a drink with separate, distinct layers of colour. Always start with the heaviest liqueur first and finish with the lightest.

right, Margarita

RUM-BASED COCKTAILS

Fifteen men on the dead man's chest
Yo-ho-ho and a bottle of rum!
Drink and the devil had done for the rest –
Yo-ho-ho and a bottle of rum!
Robert Louis Stevenson, *Treasure Island*

When you think of rum, the mental picture that it more often than not conjures up is of lazy, languorous, tropical days and slow dusks spent stretched out on a comfy hammock, cocktail conveniently to hand, as you lie back listening to the gentle lapping of the waves on a sunkissed Caribbean beach edged with palm trees gently waving in the cooling sea breeze. The association is a natural one, because the Caribbean is exactly where rum originated hundreds of years ago. Basically, it is sugar cane which has been boiled down to a rich residue called molasses. This residue is then fermented and distilled.

right, Cuba Libre

Rum comes in two main varieties – light and dark – though Cuba produces a third, called gold. In the main, light white rums like Bacardi are produced in the Spanish-speaking islands of the region, such as Puerto Rico. Because of their dryness and the fact that they have a less intense flavour than their dark counterparts, you can often substitute them for gin or vodka as a way of ringing the changes in your cocktail-making. Dark rum is the result of ageing the light spirit in oak casks – the ageing process can take from three to ten years – together with the addition of caramel colouring to the maturing alcohol.

The best dark rums come from Jamaica, though some fine rums are produced in Antigua, Barbados, the Dominican Republic, Haiti, Martinique, Trinidad and Tobago, and, of course, Cuba. Across on the South American mainland, Venezuela also boasts a particularly quaffable example. Indeed, the very best dark rums can be savoured and sipped just like a fine brandy.

PLANTER'S PUNCH

There are many, many versions of this old colonial classic, which is hardly surprising when you consider that, for centuries, the fortunes of what were then known as the West Indies depended on the sugar cane that was produced on acre upon acre of rolling plantations. Indeed, until the crash of the sugar industry in mid-19th century Tobago – where this version comes from – the expression 'as rich as a Tobago planter' had passed into the language. Some recipes call for the punch to be stirred as opposed to shaken, but, as half the fun of making cocktails comes from getting down to work with a shaker, we have chosen a version that calls for its use.

• One part dark rum
• Two parts fresh orange juice
• Juice of half a fresh lime
• Two barspoons of Grenadine
• One teaspoon caster sugar per glass
• Crushed ice
• Orange slices and pineapple chunks

Put the crushed ice in the shaker, followed by the rum, orange juice, lime juice, Grenadine and sugar, and shake well. Strain into a chilled low or highball glass, add the fruit and serve. If you like, you can top up the mixture with chilled club soda or lemonade to make what is already a long drink even longer – and, at the same time, less potent – but remember, if you do, that this must be added after the shaking. Never, ever, use fizzy mixers in a cocktail shaker.

BRASS MONKEY

For three hundred years, grizzled tars throughout the Royal Navy looked forward to that moment in the day when the order 'Up Spirits!' was given, the daily rum ration emerged and the mainbrace was well and truly spliced. This potent concoction also originated at sea, though probably in the officers' wardroom rather than the seamen's quarters: the 'brass monkey' was the name given to the metal rack on which cannonballs were stored on the mighty men-of-war of the great days of sail.

- **One part white rum**
- **One part vodka**
- **Four parts orange juice**
- **Ice cubes**
- **Orange slices**

Fill a mixing glass with the ice cubes, and pour the rum, vodka and orange juice over them. Stir carefully, then strain into a chilled highball glass and serve, garnished with the slice of orange. This tastes even better sipped slowly through a straw.

ZOMBIE

Voodoo, zombies and Black Magic all feature prominently in Caribbean folklore, especially in Haiti, where voodoo is said to have originated. Indeed, one variation on the traditional recipe given here is actually called a Zombie Voodoo to make the connection even clearer. Another variation from the U.S. calls for the knockout addition of 151-proof rum, more than enough to bring any corpse back to life. Normally, such rums are confined to desserts that call for flambéeing.

- Two parts white rum
- One part dark rum
- One part golden rum, Curaçao, or apricot brandy
- Two parts lime juice
- One part orange juice
- One part pineapple juice
- Half a part of sugar syrup
- Crushed ice
- Sprig of mint, pineapple, orange and lemon slices, and red and green cocktail cherries to garnish

Put about half a cup of crushed ice in the cocktail shaker, then add all the ingredients with the exception of the mint and the other elements of the garnish. Shake well and pour into a Collins glass – in this instance, there is no need to strain the cocktail. Stir in the mint leaves, threading the fruit together on a cocktail stick, and place in the glass.

CUBA LIBRE

Despite its name, this powerful cocktail has nothing to do with Fidel Castro, nor with the momentous revolution that finally overthrew the corrupt Battista regime, nor with the abortive Bay of Pigs landings, nor the Cuban missile crisis and the decades of confrontation with the U.S. following it. Its origins go further back than this – to the Prohibition years, when it became popular amongst the Americans who were rich enough to make it to Havana from Florida in search of a decent alcoholic tipple.

- One and a half measures of light rum
- One measure of lemon juice or juice from a freshly-squeezed lime
- Cola
- Ice cubes
- Lemon or lime slices
- Straw

You can mix this cocktail in a highball glass if you like, though, if you are making cocktails for a crowd of people, a mixing glass will be more convenient. Cover the ice cubes with the rum and juice, top up with the cola and add the lemon or lime slices. As well as a straw, it is customary to serve the cocktail with a swizzle stick.

CORKSCREW

- Three parts light rum
- One part dry vermouth
- One part peach brandy
- Ice cubes
- Slices of lime

Put the ice cubes into a cocktail shaker, add the rum, vermouth and the peach brandy and shake well. Strain into a chilled cocktail glass and add the lime slices.

PIÑA COLADA

- One part light rum
- Two parts coconut cream
- Two parts pineapple juice
- Crushed ice
- Straw
- Pineapple chunk, cocktail cherry and slither of fresh coconut to garnish

Put the ice into a cocktail shaker, followed by the rum, coconut cream and pineapple juice. Shake well and strain into a chilled Collins glass. Garnish with the fruit and serve with a straw.

BACARDI

This is not the ubiquitous Bacardi and Coke that you find in so many bars, but the genuine article.

- Two parts Bacardi rum
- One part freshly squeezed lime juice
- Two dashes of Grenadine
- Crushed ice
- Lemon slice and cocktail cherry

Put all the ingredients into a cocktail shaker – crushed ice first – and then shake together well. Strain into a Martini glass, add the lemon slice and cocktail cherry and serve. To make this cocktail even more special, try adding a further ingredient – one part of dry gin.

BEE'S KNEES

Another survivor from the long-ago days of Prohibition and the speakeasies that sprang up across the length and breadth of the U.S. The name 'speakeasy' itself probably comes from the soft tap on the door and the password that had to be murmured to the door's custodian to gain admittance. It is one of the few cocktails in which honey features as part of the brew.

- Two parts light rum
- One part freshly squeezed orange juice
- One part freshly squeezed lime juice
- A teaspoon of honey (per glass)
- Two dashes of orange bitters (per glass)
- Orange peel
- Ice cubes or crushed ice

Put the ice into the cocktail shaker, followed by the rum, juices, honey and, finally, the dashes of bitters. Shake vigorously until well blended and then strain into a chilled cocktail glass. Twist the orange peel over the glass, then drop it into the contents and serve.

CLASSIC DAIQUIRI

One of the U.S.'s favourite cocktails, the Daiquiri originated in Cuba, when American nickel miners were resourceful enough to come up with this substitute for their Bourbon-based drinks, during a national shortage of imported liquor. The name comes from the mine where the Americans worked. As with other celebrated cocktails, there are many versions of the Daiquiri, bartenders regarding it almost as a point of honour to devise their own variants. For a Derby Daiquiri, for instance, orange juice is substituted for lime, though a teaspoonful of the latter is needed for complete authenticity. The drink originated in the U.S., so it is more likely to be associated with the Kentucky Derby than its Epsom cousin.

- One part light rum
- Juice of a freshly squeezed lime or lemon
- A teaspoon of sugar or four dashes sugar syrup
- Crushed ice
- Lime slice

Fill the shaker three-quarters full with the crushed ice, and add the rum. Blend the lime juice and sugar together and add that. Then shake very thoroughly before straining into a cocktail or a Martini glass. Add the slice of lime.

BETWEEN THE SHEETS

Some mixologists contend that this should be classed as a brandy-based cocktail, but, either way, the results are delicious.

• **One part light rum**
• **One part brandy**
• **One part Cointreau or Triple Sec**
• **A dash of freshly squeezed lemon juice**
• **Lemon peel twist as garnish**
• **Ice cubes**

Half-fill the cocktail shaker with the ice cubes and then add the rum, brandy, Cointreau or Triple Sec and lemon juice. Shake well, strain and serve in a chilled cocktail glass, garnished with a twist of lemon peel.

TOM AND JERRY

Cocktails are not exclusively warm-weather drinks. Unfortunately, it is not the immortal Hollywood cartoon cat and mouse who are commemorated in this winter warmer. It is thought that its origins date back to the 1850s, when, it is believed, it was first mixed by Jerry Thomas, a famous St Louis bartender of the day. Later, the name simply got changed, probably at the same time as the original recipe was simplified substantially. The version here omits the baking soda and hot milk called for in the original. The secret, however, remains the same – a stiff mixture well imbued with alcohol and a warm coffee cup to hold the drink.

- **One egg (you will need to separate yolk and white)**
- **Caster sugar**
- **One part light rum**
- **One part brandy**
- **Grated nutmeg**
- **Boiling water**

Beat the yolk and white of the egg separately. Then combine the two, adding sufficient sugar to the mixture to stiffen it. Place the mixture in a warmed coffee cup, and add the rum, followed by the brandy and finally top up with boiling water. The boiling water takes the place of the hot milk of the original, which also had far more elaborate mixing instructions. Sprinkle the grated nutmeg over the top and serve while still piping hot.

MOJITO

A favourite of Ernest Hemingway and other Havana movers in the early 1900s, the Mojito (pronounced moe-hee-toe) is currently enjoying a reclaimed position of fame. With a reputation as the Cosmo for the more adventurous, it is no wonder that it has made a celebrity appearance on *Sex and the City*; however, it gains serious cocktail kudos in the James Bond movie *Die Another Day* when the bikini-clad Halle Berry emerges from the sea and Bond utters simply 'Mojito?'.

- Handful of mint sprigs (about 5)
- Two parts white rum
- One part lime juice
- Two dashes of sugar syrup
- Splash of club soda
- Crushed ice

Place the mint in the bottom of a highball glass, then pour over the rum, lime juice and sugar syrup. Using a barspoon, pound the ingredients to release the flavour of the mint. Add the ice, stir well; then top up with the club soda. Give it all another good stir and serve immediately.

ACAPULCO

In the words of the old song, 'south of the Border, down Mexico way' lies the lazy resort town of Acapulco. This deliciously sinful cocktail should bring back happy memories to anyone who has visited the bars and restaurants of the resort. The only thing left to work out is what to do with the redundant egg yolk not used in this recipe!

• **One part light rum**
• **A tablespoonful of lime juice**
• **One-and-a-half teaspoons Triple Sec**
• **One teaspoon caster sugar**
• **A single egg white**
• **Crushed ice and ice cubes**
• **Mint sprigs for garnish**

Fill a cocktail shaker half-full of crushed ice. Add the other ingredients, shake thoroughly and then strain into old-fashioned glasses over the ice cubes, allowing for a couple of ice cubes a glass. Garnish each glass with a sprig of mint.

MARY PICKFORD

There is at least one other cocktail that claims to take its name from Mary Pickford, silent-screen film goddess, wife of Douglas Fairbanks Senior, and hailed as 'America's sweetheart' in her day. Whichever of the brews she herself favoured, she was breaking the law, for this was the time of the 'noble experiment' of Prohibition.

- **One part light rum**
- **One part pineapple juice**
- **A dash of Grenadine**
- **A dash of Maraschino**
- **Crushed ice**

Half-fill a cocktail shaker with crushed ice, add all the other ingredients and shake thoroughly. Strain into a Martini glass.

right, Mary Pickford

MAI TAI

Despite its Oriental-sounding name, full of Eastern promise, this tropical cooler also hails from the Caribbean. It is ideal drinking for a sultry, humid summer evening.

- **One part light rum**
- **One part dark rum**
- **One part Triple Sec, Cointreau or apricot brandy**
- **A dash of Grenadine**
- **A teaspoon of lime juice**
- **Three parts of freshly squeezed chilled orange juice**
- **Three part of freshly squeezed chilled pineapple juice**
- **Crushed ice**
- **Slice of pineapple and mint sprig for garnish**

Half-fill a cocktail shaker with ice. Then add the white and dark rum, the Triple Sec and the Grenadine, working in that order. Follow with the fruit juices. Shake the mixture well until the outside of the shaker starts to feel cold, then strain over the crushed ice into large old-fashioned glasses, garnish and serve. For that extra kick, top with a dash of 151-proof rum.

LONG ISLAND TEA

'Long Island' and 'Tea' both carry with them a sobering suggestion of Yankee responsibility and respectability. In fact, though deceptively thirst-quenching, this long drink can be both potent and intoxicating, though the sting in the tail is often concealed by the topping-up of cola called for in the recipe. If you are braver than the average, you could always cut right back on the cola.

- **One part light rum**
- **One part vodka**
- **One part gin**
- **One part Tequila**
- **Juice of half a lemon**
- **Two parts cold Earl Grey tea**
- **Chilled cola, to taste**
- **Cracked ice and ice cubes**
- **Lemon (or lime) wedges and a sprig of mint to garnish**

Put plenty of cracked ice into a bar glass and add the rum, vodka, gin and tequila, followed by the lemon juice. Then, add enough Earl Grey tea to make a good long drink, stir well and then strain into a highball glass, filled with ice cubes and lemon wedges. Top up with the chilled cola and garnish.

BRANDY-BASED COCKTAILS

Claret is the liquor for boys; port for men; but he who aspires to be a hero must drink brandy.
James Boswell, *The Life of Samuel Johnson*

The good doctor did not get it far wrong – brandy, as a liquor, stands in a class all of its own. There are many fine varieties, the best-known being France's heady cognacs and rich armagnacs, though there are fine Californian, Greek, German, Australian and South African varieties. All of these are distilled from wine. In addition, there are apple brandies – Calvados, which by law must come from Normandy, being the classic example – which are distilled from cider made from apples, as well as a variety of brandies distilled from other fruits. Those made from cherries, for instance, are called Kirsch or Kirschwasser, while those made from pears are Poire, and from raspberries, Framboise. Collectively, they are known as *eaux-de-vie*. Fruit-flavoured brandies are brandy-based liqueurs, flavoured with peaches, apricots, blackberries, and other fruits. All of them are surprisingly versatile and are an essential part of the mixologist's armoury.

right, Brandy Sour

Vintage brandy, however, is another matter entirely. This is a great drink in its own right and tastes much better on its own, completely unadulterated. If you are lucky enough ever to get a taste of the real Napoleon – or, indeed, any fine vintage brandy – you will immediately see why it would be positively sinful to waste it in a cocktail. So, for the purposes of this book, we suggest that you stick to a good non-vintage tipple unless a specific variety is called for.

HINE

SIGNATURE
PETITE CHAMPAGNE

COGNAC

B&B

Here's a classic combination to get you started – the brandy in the recipe is the natural companion for the Benedictine, a herb-flavoured brandy liqueur, the exact constituents of which have been a closely-guarded secret ever since the Benedictine monks of northern France produced the first Benedictine back in those medieval days of yore. If your fancy is for a longer, cooler drink, simply add some club soda and you will have a B&B Collins.

• **One part brandy**
• **One part Benedictine**
• **Ice cubes**
• **A twist of lime to garnish**

Place a few ice cubes in a bar glass and add the brandy and the Benedictine. Stir well and then strain the mixture into a cocktail glass. Garnish with the lime twist and serve. As an alternative, you could try making the cocktail directly in a cordial glass, the trick here being to physically float the brandy on top of the Benedictine.

BRANDY ALEXANDER

History does not record what exactly was in the 'four Alexandra cocktails' which the arch-aesthete Anthony Blanche ordered as a pre-dinner aperitif in Evelyn Waugh's immortal novel *Brideshead Revisited*. It could have been this cocktail, or its gin-based cousin. What we do have Waugh's assurance for is that what was ordered, to the scandal of every eye in the bar of Oxford's George Hotel, was sweet, creamy and delicious.

- One part brandy
- One part Crème de Cacao
- One part thick, sweetened cream
- Crushed ice or ice cubes

Put enough ice in a cocktail shaker to fill it a quarter full, then add all the ingredients and shake thoroughly. Strain into a chilled Martini glass. Traditionally, that's all there is to it, but you could garnish with a sprinkling of grated nutmeg if you like. To ring the changes, try substituting gin for the brandy and Crème de Menthe for the Crème de Cacao.

CHARLESTON

Anyone who aspired to be anyone among the jazz-crazy Bright Young Things and Flaming Youth of the 20s had to master the energetic Charleston before taking confidently to the dance floor. This potent cocktail may well have helped them along the way – after all, you never knew when you might be dancing 'with a man who's danced with a girl who's danced with the Prince of Wales'!

- **One part cherry brandy**
- **One part orange liqueur**
- **Chilled lemonade to taste**
- **Ice cubes**

Pour the cherry brandy and the orange liqueur into a bar glass. Stir well, then fill a highball glass with ice cubes and pour the cocktail directly over them. There is no need to use a strainer in this instance. Top up with the chilled lemonade to taste, then serve.

AU REVOIR

Take a look at the ingredients listed below, and you can probably guess how and why this cocktail was given its name. Sloe gin is something you can easily make for yourself, provided that you have the time and patience to let the brew ripen and mature. For a little extra variety, you can replace the brandy with the apricot variety and eliminate the egg white, in which case you will be drinking a Charlie Chaplin, as this variant was christened in the.U.S. Maybe this was something to do with the way people walked after downing a tincture or two.

• **One part brandy**
• **One part sloe gin**
• **Juice of one lemon**
• **One egg white**
• **Ice cubes**
• **Lemon slice to garnish**

Place four ice cubes in a cocktail shaker and add the brandy, sloe gin, lemon juice and, finally, the egg white. It is important to follow this order for the best result. Shake well and strain into a Martini glass. Garnish with a lemon slice before serving.

EGGNOG

In times past, this was an English country house favourite, served on Christmas mornings as a post-church warmer to drive out the chill. There are many possible variations on the basic recipe: you could add some Madeira or Crème de Cacao to it, use apricot brandy and Triple Sec instead of the conventional cognac and rum, substitute whisky for the brandy and change the dark rum for the light variety, or, for a really sinful version, use some whipped double cream as well as the milk. Traditionally, though, whichever mixture you decide on, you should sprinkle some nutmeg on the top of each serving.

- One part brandy
- One part dark rum
- One egg
- A dash of sugar syrup
- Five parts warm milk
- A little grated nutmeg

Shake the brandy, rum, egg and syrup in a cocktail shaker. This should be done vigorously, so that the ingredients all blend into a deliciously creamy mixture. Strain into a highball glass, add the milk – this can be cold milk, if preferred – and stir gently. Garnish with grated nutmeg before serving.

BRANDY CLASSIC

There are literally dozens of cocktails simply entitled Brandy Cocktails – it's almost as if bartenders around the world have all vied with one another to create their own individual versions. That's as it should be, but the recipe given here is for an accepted standard. For variety, substitute two dashes of Grenadine for the bitters, cut out the sugar syrup, and add some freshly squeezed lemon juice (the juice from one lemon should be enough). The result is known as a Brandy Gump – no relation, as far as is known, to Forrest Gump!

- Two parts brandy
- A quarter teaspoon of sugar syrup
- Two dashes of Angostura bitters
- Ice cubes
- Lemon twist as garnish

Stir the ingredients, along with the ice cubes, in a mixing glass and then strain into a cocktail glass. Serve garnished with a twist of lemon peel threaded on a cocktail stick.

SIDECAR

In the early years of the century, Harry's New York Bar, confusingly enough located in Paris, was the European headquarters of the cocktail boom. This well-loved classic was apparently first mixed there in 1911: it was devised for a somewhat eccentric customer, who arrived for his daily visits to the bar in the sidecar of a chauffeur-driven motorcycle. Some people say you can simply mix a Sidecar in a mixing glass, but we think that it tastes better shaken.

- One part brandy
- One part Triple Sec
- One part freshly squeezed lemon juice
- Crushed ice or ice cubes
- Orange or lemon slice to garnish

Fill a cocktail shaker three-quarters full with ice, add the brandy, Triple Sec and lemon juice, shake well and strain into a frosted Martini glass with a sugared rim. Garnish and serve.

BRANDY SOUR

In the cocktail world, sours come in all kinds of varieties – as well as brandy sours, there are rum, whisky, gin, vodka and tequila equivalents. All of them should taste tart and lemony: they are extremely refreshing on a hot, humid day. If the recipe here sounds too sour for your taste, add a part of freshly squeezed orange juice to it, plus a further part of brandy.

• Two parts brandy
• One part freshly squeezed lemon juice
• A teaspoon of caster sugar per glass
• Ice cubes

Half-fill a cocktail shaker with ice cubes and pour in the brandy and the lemon juice. Add sugar to taste, then shake well. Strain into a Martini glass – or a sour glass, if you have one to hand – garnish and serve.

Horse's Neck

HORSE'S NECK

To give this an extra kick, you can substitute whisky for the brandy. Again, this is not an elaborate cocktail – you do not even need a mixing glass.

- **One part brandy**
- **Ginger ale**
- **Ice cubes**
- **A spiral of lemon peel to garnish**

Place a couple of ice cubes in a highball glass and pour the brandy over them. Garnish with the spiral of lemon peel and then top up the glass with ginger ale to taste.

STINGER

Like a scorpion, the sting here is very much in the tail. A couple of Stingers is the perfect prescription for getting guests to relax, unwind and start enjoying themselves. In the days of Prohibition, Stingers were served straight up, but the tendency now is to serve them on the rocks. Traditionally, they were also served ungarnished, though you can add a sprig of mint if you like.

- **Three parts brandy**
- **One part white Crème de Menthe**
- **Ice cubes**
- **Sprig of mint as garnish (optional)**

Half-fill a cocktail shaker with ice cubes. Add the brandy and the Crème de Menthe, shake well and strain into a chilled Martini or cocktail glass.

GRENADIER

No, this cocktail's name does not derive from the fact that it started off as a favourite tipple in the officers' mess of one of Britain's most prestigious Guards' regiments. It comes, we think, from the Grenadine in the drink. This is one of the few cocktails where the specific use of Cognac is often stipulated. Traditionally the cocktail is always served ungarnished.

- **One part cognac**
- **One part ginger wine**
- **Three dashes of Grenadine**
- **Ice cubes**

Half-fill a cocktail shaker with ice cubes and then add the cognac and ginger wine, finishing off with the dashes of Grenadine. Shake well, strain into a Martini or cocktail glass and serve.

COFFEE

No one knows where, when, how or why this cocktail got its name: the only coffee that is involved in its mixing are the beans that are applied as the garnish before the cocktail is served. Certainly, there is no long-standing tradition involved as far as we are aware – unlike the Italian tradition of garnishing a straight Sambucca with coffee beans, flaming the drink and letting it burn until the beans just start dissolving enough to begin colouring the liquid.

- **Three parts brandy**
- **One part port**
- **An egg yolk**
- **A teaspoon of sugar per glass**
- **Ice cubes**
- **Coffee beans or grated nutmeg to garnish**

Half-fill the shaker with ice cubes, add the brandy, port, sugar and egg yolk and shake well. Strain into an old-fashioned glass. Either float a few roasted coffee beans on top to garnish, or, as they do in the U.S., serve with a sprinkling of nutmeg over the top.

HARVARD

As you'd expect, this cocktail takes its name from one of the U.S.'s most prestigious Ivy League universities – though tradition does not record whether it was first devised on campus, or in the comfortable bar of the Harvard Club in New York. Not to be outdone, Yale, Harvard's great rival, has its own cocktail as well. Obviously, this simply could not be brandy-based, so the Yale 'bulldogs' went for their own potent mixture of three parts gin to one of dry vermouth, a dash of bitters and, to top things off, a spoonful or two of Blue Curaçao. Getting back to the Harvard cocktail, mixologists differ as to whether it is best to mix it or shake it, but, on the whole, the shakers have it.

• **One part brandy**
• **A dash of sugar syrup**
• **One part sweet vermouth**
• **A dash of Angostura bitters**
• **Ice cubes**
• **A slice of lemon to garnish**

Half-fill the shaker with ice cubes and then add the brandy, sugar syrup and sweet vermouth. Shake well. Before straining into a cocktail glass evenly, coat the sides of the glass with the dash of bitters. Then strain the cocktail into the glass and garnish with lemon.

BOSOM CARESSER

You've already met Between the Sheets; now, here comes another seductive favourite in the form of the Bosom Caresser. It's interesting just how many of these there are – Kiss the Boys Goodbye, Lady Be Good and the Widow's Kiss are just three further examples. Kiss the Boys Goodbye makes good use of a sloe gin and brandy combination, while Lady Be Good relies on brandy, white Crème de Menthe and sweet vermouth for its punch. And the widow who first indulged in a Widow's Kiss – a shake of Calvados, Yellow Chartreuse and Benedictine – must have been a very merry widow indeed.

- **Three dashes of Grenadine**
- **One egg yolk**
- **Two parts brandy**
- **One part orange Curaçao**
- **Ice cubes**

Half-fill the shaker with the ice cubes, add the Grenadine, egg yolk, brandy and Curaçao, and shake well. Strain into a chilled champagne saucer and serve.

BLACKSMITH COCKTAIL

Sweating away all day over a fiery forge must have been hard work, so perhaps it isn't surprising that an Irish blacksmith a hundred years or so back is credited with coming up with the interesting idea of the original Blacksmith. This was a stiff mix of half a pint of Guinness and the same amount of barley wine, which the blacksmith saw as the ideal way of ending a day of labour. The cocktail equivalent, though slightly more sophisticated, is equally delightful.

- **One part brandy**
- **One part Drambuie**
- **One part Crème de Café**
- **Ice cubes**

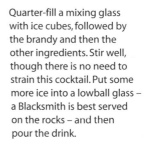

Quarter-fill a mixing glass with ice cubes, followed by the brandy and then the other ingredients. Stir well, though there is no need to strain this cocktail. Put some more ice into a lowball glass – a Blacksmith is best served on the rocks – and then pour the drink.

BOMBAY COCKTAIL

For most people, Bombay, in drinking terms, is associated with Bombay gin, but this powerful cocktail must have raised a few eyebrows among the sahibs and the memsahibs of the heady days of the British Raj – that is, assuming the cocktail ever made it back to the city after which it was named. If it did, it may well have been given short shrift by the staunch stiff-upper-lip upholders of the pink-gin-now-that-the-sun's-over-the-yardarm tradition.

- Two parts brandy
- One part dry vermouth
- One part sweet vermouth
- A quarter spoonful Anisette
- A half spoonful Triple Sec
- Crushed ice or ice cubes

Quarter-fill a mixing glass with ice, add the ingredients and stir well. Strain into a chilled Martini glass. Serve ungarnished.

GIN-BASED COCKTAILS

Don't tell my mother I'm living in sin
Don't let the old folks know
Don't tell my twin that I breakfast on gin
He'd never survive the blow.
Sir Alan Herbert, *Laughing Anne*

Probably the most frequently-used cocktail base of them all, gin, is officially classed as a 'neutral, rectified spirit, distilled from any grain, potato or beet and flavoured with juniper'. The trick lies in the way that the raw spirit is reprocessed and redistilled. For, although it is the most important extra ingredient, juniper is not the only one to be added. It is accompanied by a mixed assortment of various other herbs and spices. Most gin-producers have their own closely-guarded recipes.

The drink was said to be invented by a Dutch doctor some three hundred or more years ago, when he was looking for something new in the way of medicine to

right, Dry Martini

try out on his patients. Whether this charming tale is true or not – certainly the first gin did come from the Netherlands – what is known is that its popularity grew quickly. Regardless of whether or not it actually had any curative powers, it made people feel better to drink it, which, along with its cheapness, helped to make it the favourite spirit among Britain's poor in the country's towns and cities in the 18th century. The artist William Hogarth immortalized this facet of gin-drinking in his Gin Lane series of caricatures: in this infamous lane, the promise was 'Drunk for a penny, dead drunk for twopence'.

Indeed, as far as high society was concerned, gin only became a respectable tipple with the coming of cocktails. The Martini – the most celebrated cocktail of them all – is firmly gin-based in its traditional form, though there have been Johnny-come-lately attempts to substitute vodka for gin in some more modern recipes. Other gin-based classics include the Gin Sling, Tom Collins, White Lady and the Bronx.

PINK LADY

A Pink Lady is, as its name implies, a deliciously decadent relation of the White Lady that fully merits the garnish it is traditionally given. The drink tastes just as wonderful as it looks. Some more puritanical recipes leave out the cream, but, as the old adage goes, 'why spoil the ship for a ha'porth of tar'?

- **One part gin**
- **One part Calvados**
- **One part lime juice**
- **Half an egg white per glass**
- **A teaspoon of cream per glass**
- **Five dashes of Grenadine**
- **Ice cubes or crushed ice**
- **Slice of lime garnish**

Half-fill a cocktail shaker with ice, add the gin, Calvados, lime juice, cream and egg white and shake thoroughly. Shake well and strain into a previously sugar-frosted cocktail glass. Garnish with the lime slice.

White Lady

WHITE LADY

This all-time cocktail classic was first created in Harry's New York Bar, Paris, though, sadly, no one appears to know who the original white lady was nor how this delicious cocktail got its name. Again, there are variations from which to choose. Some bartenders of the old school prefer to mix the cocktail, rather than shake it, while others advocate the use of a tablespoonful of single cream and another of caster sugar, rather than the egg white, and leaving out the lemon juice. What goes without saying is that, despite those who claim that vodka can be substituted for the gin, the latter and the Cointreau are both totally essential. Otherwise, the choice is yours!

- **One part gin**
- **One part Cointreau**
- **One part freshly squeezed lemon juice**
- **Half an egg white**
- **Ice cubes**

Quarter-fill a cocktail shaker with ice cubes, followed by the gin, Cointreau, lemon juice and egg white. Shake thoroughly and then strain into a Martini glass. The tradition is to serve a White Lady unadorned and ungarnished.

GIN SOUR

All sours are tart, lemony, highly-concentrated cocktails. As with many other famous brews, the basic recipe has spawned many variations, as individual bartenders around the world have added their own individual touches to it. As well as the standard classic, we are giving a Canadian variant on it, too, as a demonstration of how easy it would be for you to experiment to give your own cocktails a personal twist in terms of flavour and savour.

- Three parts gin
- One part freshly squeezed lemon juice
- Three dashes sugar syrup
- Ice cubes
- Half-slice of lemon and cocktail cherry to garnish
- A half egg white (for the Canadian variation)
- Ice cubes
- Half slice of lemon to garnish

Half-fill a cocktail shaker with the ice cubes, add the lemon juice, sugar syrup, egg white (if wanted) and the gin. Shake well and then strain into a sour glass. Garnish with the half slice of lemon, or for the Canadian variation add the slice of lemon to the drink.

GIN SMASH

This is an extremely refreshing cocktail that is simplicity itself to shake and make. To vary the flavour, try substituting other kinds of mint for the fresh peppermint specified here – apple mint is just one example of another type of mint you could try.

- A lump of sugar
- Soda water
- Four sprigs of fresh peppermint
- Two parts gin
- Cracked ice
- Slice of lemon and mint to garnish

Muddle the sugar with the soda water and the mint in a cocktail shaker and then half-fill the shaker with the cracked ice. Add the gin and shake well – the ingredients must be shaken enough to ensure that the flavour of the mint gets into the gin. Strain into a glass and add the lemon slice and some mint.

GIN FIZZ

Nothing could be simpler to make than this
deliciously long cocktail. Nothing is more refreshing
to sip slowly on a hot, sultry day. As an alternative, try
substituting ice cold ginger beer for the soda water.
Another variation involves adding either the white or
yolk of an egg to the mixture in the shaker just before
you start shaking. Yum!

• Two parts gin
• Juice of half a lemon
• One egg
• A teaspoonful of caster sugar per glass
• Soda water or ginger beer
• Ice cubes
• An orange slice to garnish

Shake the gin, sugar, egg and the lemon juice with
some of the ice, strain into a highball glass over
two ice cubes, and top off with the soda water or
ginger beer. Stir gently. Serve garnished with a slice
of orange.

CARDINALE

This is a Martini with a difference, first devised and mixed in Harry's Bar in Venice. Today, it has become such a popular choice that it is widely available in ready-mixed bottled form.

- **Six parts gin**
- **One part dry vermouth**
- **Three parts Campari**
- **Ice cubes**

Mix the gin, vermouth and Campari together in a bar glass, together with a few ice cubes. Strain the result into a Martini glass and serve ungarnished.

NEGRONI

In Anthony Powell's *A Dance to the Music of Time* novels, this cocktail – a long-time traditional Italian favourite – features in *Temporary Kings*, where it is ordered 'with an urgent request for plenty of gin'. Though the recipe given here is probably not as potent as the one Powell's character had in mind, half-close your eyes, imagine sipping one of these in a bar alongside the Grand Canal in Venice, sit back and enjoy.

- One part gin
- One part sweet vermouth
- One part Campari
- Crushed ice and ice cubes
- Soda water (if liked)
- Lemon rind
- Half a slice of orange to garnish

Shake the ingredients together thoroughly with plenty of crushed ice and strain into a cocktail glass (some prefer to use an old-fashioned glass) over the ice cubes. Twist a strip of lemon rind over the surface of the drink – this gives it added zest – then drop the rind into the glass. Garnish with the half slice of orange. If you like, top up with a splash of soda before serving – but no more than a splash.

CARUSO

With apologies to the 'three tenors' of today, Enrico Caruso was undoubtedly the greatest tenor operatic star of the entire 20th century, with a unique quality in his voice that even the scratchy 78s of the early days of gramophone recording managed to capture. History, however, does not record whether or not this cocktail contributed to giving Caruso's voice its unique depth of timbre and warmth of tone.

- Two parts gin
- One part dry vermouth
- Half a part green Crème de Menthe
- Ice cubes

Put some ice cubes into a mixing glass, add the gin and then the other ingredients and stir well. Strain into a Martini glass and serve ungarnished.

left, Enrico Caruso

SNOWBALL

This drink tastes as good as it looks. It's simple and quick as well – a far better way of passing the time than getting frozen playing out in the snow.

- **Two parts gin**
- **One part Anisette**
- **A tablespoon of single cream**
- **Ice cubes**

Quarter-fill a cocktail shaker with the ice cubes, add the ingredients and shake well. Strain into a Martini glass and serve just as it is, without any garnishing or other decoration. Some bartenders, however, suggest sipping the concoction slowly through a straw.

BLUE DEVIL

The Blue Curaçao gives this stylish cocktail its striking colour – the gin its punch. You can vary the recipe by reducing the amount of gin by one part, but making up for this by using Cointreau and substituting lime juice for the lemon juice. Bartenders call this variant a Blue Arrow, which, unlike its equally potent relation, is traditionally served ungarnished, while the Blue Curaçao should be added last.

- **Three parts gin**
- **One part Blue Curaçao**
- **One part lemon juice**
- **One tablespoon Maraschino (optional)**
- **Ice cubes or crushed ice**
- **Slice of lemon as garnish**

Quarter-fill a cocktail shaker with the ice cubes or crushed ice, then add the ingredients, following the order given, and shake vigorously. Strain into a Martini glass, garnish and serve. For a Blue Arrow, chill the glass first.

Tom Collins

Atlantic substituted a drier gin, called Old Tom, for the Dutch-style original, and, in its altered form, the drink instantly caught on with the cocktail crowd. The name changed accordingly. So, too, did the amounts of gin used in the making of the drink – depending on personal taste, you can vary the amount of gin upwards or downwards by one part.

• A dash or two of sugar syrup
• Two parts gin
• One part lemon juice
• Soda water
• Ice cubes
• A slice of lemon and a few mint leaves as garnish

Half-fill a cocktail shaker with ice cubes, add the sugar syrup (you can use a teaspoonful of caster sugar if there is no sugar syrup to hand), the gin, and the lemon juice, shake well and strain into a chilled Collins or highball glass. Top up with the soda water, stir gently, garnish and serve. If you like the drink really cold, add an ice cube or two before you top up with the soda water.

SINGAPORE SLING

Cocktail history does not record whether the famous Long Bar at Raffles Hotel in Singapore was where this exotic sling first saw the light of day, but, whatever its origins, it soon became a firm classic. The writers Somerset Maugham and Joseph Conrad both singled it out as one of their favourite pre-prandial cocktail tipples.

• One part freshly squeezed lemon juice
• A tablespoonful of caster sugar
• Two parts gin
• Soda water
• One part cherry brandy
• Ice cubes
• Seasonal fresh fruit as garnish

Put four or five ice cubes into a cocktail shaker, and add the lemon juice, the sugar and the gin. Shake well and strain into a Collins or highball glass. Add more ice, top up with the soda water and then float the cherry brandy on the top, finally garnishing with seasonal fresh fruit. The cocktail tastes best sipped through straws.

right, Somerset Maugham

- Two parts gin
- One part lime juice
- Ice cubes
- Lime wheel to garnish

Quarter-fill a cocktail shaker with ice cubes, then add the gin and the lime juice, and shake thoroughly. Strain into a chilled cocktail glass, garnished with a lime wheel.

GREEN DEVIL

Green demons, dragons and hornets also feature in the cocktail repertoire, though, for most of us, a Green Devil is more than enough to handle.

- Two parts gin
- One part lime juice
- Two parts green Crème de Menthe
- Ice cubes

Quarter-fill a cocktail shaker with ice cubes, add the gin, lime juice and Crème de Menthe, shake well and then strain into an old-fashioned glass over a couple of ice cubes. Crush the mint leaves to release their aroma and then simply drop them into the drink.

GIBSON

Some mixologists claim that the Gibson – so-called because it was first mixed for the artist Charles Dana Gibson by Charley Connolly, bartender at the Player's Club, New York, during Prohibition – is nothing more than a refined version of a Dry or Extra Dry Martini. Most recipes say that a single cocktail onion is sufficient garnish, but, somehow, one is never enough. For an extra touch of authenticity, rub the glass with a clove of garlic before pouring.

• Three parts gin
• One part extra dry sherry
• Ice cubes
• Three pearl cocktail onions as garnish

Place a couple of ice cubes in a bar glass, add the sherry followed by the gin and stir together well. Strain into a chilled old-fashioned glass, and garnish with the cocktail onions.

Pink Gin

PINK GIN

Probably the most pukka of all Britain's contributions to the world's cocktail heritage, the pink gin was the standard wardroom tipple for generations of Royal Navy officers, following the drink's invention, which was probably in the West Indies. Although some modern-day cads believe that ice and a twist of lemon improves on the original, no true sahib would dream of tampering with the classic recipe.

• Two dashes of Angostura bitters
• Gin

Put the bitters into a cocktail glass and then rotate the glass until it is liberally coated with the bitters. Fill with neat gin.

VODKA-BASED COCKTAILS

It was my Uncle George who discovered that alcohol was a food well in advance of modern medical thought.
P. G. Wodehouse, *The Inimitable Jeeves*

Whether or not Bertie Wooster's Uncle George was a hardened vodka-imbiber, we shall never know – though we do know that Jeeves' celebrated pick-me-up made him feel better than he had done for years. But that's not surprising. Among spirits, vodka is the one with the most flexibility. According to some mixologists, it is pretty near the ideal cocktail spirit because it is colourless, tasteless and odourless, so adding the requisite alcoholic bite without masking the flavour of the other ingredients in the drink. Nor, unlike other spirits, does it leave any lingering tell-tale smell behind on the breath.

right, Bloody Mary

Small wonder that the Russians, who first distilled the drink many centuries ago, derived its name from the words Zhiznennia voda, which means 'water of life'. Certainly, this is what the Russian Army believed both in Tsarist and in modern times: their daily vodka ration helped many Russian soldiers to keep fighting on when, by all conventional military opinion, their position was hopeless. In the West, vodka started its rise to true popularity after the Second World War; nowadays, alongside the traditional plain spirit – the best brands come from Russia, Poland and Sweden – lemon, pepper, lime, mint, cherry, bison-grass, honey and many other specially-flavoured vodkas have hit the bars in response to ever-growing demand. The best way to drink these is neat in a small glass. The vodka should be chilled almost to freezing, and, if you are brave enough, tossed straight back in one single gulp. However, it could prove cripplingly expensive to practice the final rite of the Russian tradition – after a toast, smashing the emptied glasses into a fireplace, or on to the floor.

BLOODY MARY

The original Bloody Mary, it is said, was devised by Fernard Petiot, a noted mixologist of the 20s, when he was working in Harry's New York Bar, Paris. Petiot named the cocktail after Mary Pickford, nicknamed 'America's sweetheart' and unquestionably one of the greatest Hollywood leading ladies of the day. History, it seems, does not record what she thought of the name, though she certainly had the reputation of being difficult to handle. Nowadays, there are literally hundreds of versions of what has become probably the world's most popular vodka-based cocktail.

- **Two parts vodka**
- **Six parts tomato juice**
- **Two dashes Worcestershire sauce**
- **One dash lemon juice**
- **One dash of Tabasco sauce**
- **Pinch of celery salt**
- **Pinch of white or black pepper**
- **Ice cubes**
- **A stick of celery as garnish**

Place four or five ice cubes in a cocktail shaker, and add the vodka, tomato juice, Worcestershire sauce, lemon juice,

Tabasco sauce and the celery salt. Shake well and strain into a highball glass over some more ice cubes. Garnish with the celery stick, lightly dusting the drink's surface with the pepper. This is an old Russian tradition where vodka is concerned: in the days when the spirit was not always that well-distilled and pure, the pepper dusting forced potentially harmful sisal oil to the surface. At least, that's how Ian Fleming's James Bond explained the affectation to 'M' over dinner in Moonraker.

For once, you should not be afraid to vary things to suit your particular palate – if you like your Bloody Mary spicier, increase the amounts of Worcestershire sauce and the Tabasco. For a complete change, try substituting tequila for the vodka, when you will end up with what South Americans call a Bloody Maria. Or, if you use a clam juice and tomato juice mixture, you'll be drinking a Bloody Muddle!

BLUE LAGOON

A cooling ice-blue summer drink that's a real refresher when sundowner time comes around. For an even longer version, you could omit the pineapple juice and simply top up with some lemonade – it's the vodka and the blue Curaçao that give this drink its delicious kick. Some recipes suggest adding a couple of dashes of Green Chartreuse, but our feeling is that this detracts from the classic simplicity of the long-established original.

• Three parts vodka
• One part Blue Curaçao
• Three parts pineapple juice or lemonade
• Crushed ice
• Maraschino cherry to garnish

Quarter-fill a cocktail shaker with crushed ice, and add the vodka, Blue Curaçao and pineapple juice. Shake thoroughly and strain into either a low or highball glass. Garnish with the slice of pineapple and the cocktail cherry before serving.

MOSCOW MULE

Cocktail tradition has it that this curiously-titled cocktail – apparently Los Angeles barman John Martin named it after Smirnoff's Moscow factory when he devised the drink in 1947 – should, strictly, be mixed in and drunk from a copper mug, but this is a completely dispensable affectation. Today, the cocktail actually comes ready-mixed in bottles, but there is no substitute for shaking your own home version. Ring the changes by substituting gin for the vodka – the result is just as deliciously thirst-quenching.

• Three parts vodka
• Three dashes Angostura bitters
• One part lemon or lime juice
• Ginger beer
• Ice cubes
• Lime wedges to garnish

Pour the vodka, bitters and lemon juice over a couple of ice cubes in a chilled highball glass. Add the lime wedges to the drink and top up with the ginger beer. Stir and serve.

COSMOPOLITAN

Affectionately known as the 'Cosmo', this cocktail has become the thirty-something's best friend. Its fame rocketed with the success of the show *Sex and the City*, and it has never looked back…! In one small cocktail glass, the Cosmo encapsulates all that is New York sex appeal and urban glamour.

- **One and a half parts vodka**
- **One part Cointreau**
- **One part cranberry juice**
- **Dash of lime juice**
- **Ice cubes or cracked ice**
- **Fresh berries or lime twist to garnish**

Half-fill a cocktail shaker with ice, then add the vodka, Cointreau, cranberry juice and lime juice – shake thoroughly. Strain the liquid into a frosted Martini or cocktail glass and garnish with the fresh berries or lime twist. Serve.

ORANGE COSMO

- One and a half parts vodka
- One part Grand Marnier
- One part cranberry juice
- Dash of lime juice
- Ice cubes or cracked ice
- Orange twist to garnish

Add enough ice to half-fill a cocktail shaker, then add all the ingredients except the garnish. Shake vigorously, then strain into a chilled martini or cocktail glass, and serve garnished with the orange twist.

THE BLACK MARBLE

One thing that the overwhelming majority of tried-and-tested cocktail classics have in common is their straightforwardness. The Black Marble, which some liken to a good Dry Martini, is no exception to the precept. The better the quality of the vodka, the better the drink tastes.

- One part vodka
- Ice cubes
- A single large black olive
- A lemon or orange wedge to garnish

Fill a low ball glass three-quarters full of ice cubes. Place the black olive in the centre of the pile of cubes and pour the vodka over it. Garnish with the fruit slice and serve.

SALTY DOG

In the U.S.A., this started off life as a gin-based cocktail, but now it is well on the way to becoming an established vodka classic. It gets its name from the ring of salt that you frost around the cocktail glass.

• **Four parts vodka**
• **One part unsweetened grapefruit juice (fresh is best)**
• **Ice cubes**
• **Salt for frosting**

Put four ice cubes in a cocktail shaker, add the vodka and the grapefruit juice and shake thoroughly. Frost the rim of a well-chilled Martini glass with salt, then strain the cocktail into the glass. The alternative is simply to pour the ingredients, not forgetting the salt, over ice cubes into a highball glass, but, somehow, this sounds like far less fun.

HARVEY WALLBANGER

The story behind how this cocktail got its name sounds so unlikely that it may even be true. According to bar-room gossip, a Californian surfer called Harvey, having been eliminated from an important championship, went on a binge to drown his sorrows. Having downed excessive amounts of vodka and Galliano, he ended up by banging his head in frustration against the bar wall. It's up to you to judge!

- Two parts vodka
- One part Galliano
- Fresh orange juice
- Ice cubes
- A slice of orange to garnish

Quarter-fill a cocktail shaker with ice cubes and add the vodka and orange juice. Shake well and strain into a highball glass over two more ice cubes. Gently float the Galliano on top, then garnish with the slice of orange. It is traditional to serve this drink with a stirrer and to sip it through a straw.

VODKA MARTINI

It was Ian Fleming's James Bond, who first introduced vodka into the Martini, at least in the fictional world.

The original recipe (see The Classic Martini) married vodka with gin: this version is more straightforward. If you want to avoid trouble, remember the immortal Bond precept: 'Shaken, not stirred'.

• **Two parts vodka**
• **One part dry vermouth**
• **Ice cubes**
• **A green olive or twist of lemon rind**

Half-fill a cocktail shaker with the ice cubes, add the vodka and vermouth and shake well. Strain the mixture into a chilled cocktail glass, in which you have pre-placed a green olive or twist of lemon. Serve.

LONG VODKA

Add a few drops of Angostura bitters to a chilled highball glass and swirl around so that the whole glass is coated. Empty out excess. Add ice, the remaining ingredients and garnish.

- Angostura bitters
- One shot vodka
- Generous dash of lime
- Top up to fill level with lemonade
- Serve with a straw and a slice or twist of lime

SOVIET

The Soviet Union might have come to an end, but this classy little drink is still full of Russian style and punch. Also, this cocktail is the nearest we could get to finding a Communist counterpart for the languorous White Russian that featured earlier in this chapter. Be careful – it is stronger than it might appear on the surface.

- Three parts vodka
- One part Amontillado sherry
- One part dry vermouth
- Ice cubes
- A twist of lemon peel

Quarter-fill a cocktail shaker with ice cubes, add the vodka, sherry and vermouth and shake well. Strain over more ice cubes into an old-fashioned glass. Add a twist of lemon peel before serving.

SCREWDRIVER

Another inspiration from Los Angeles mixologist John Martin, inventor of the equally celebrated Moscow Mule. Again, you can simply mix this in a highball glass over ice cubes, but, like John Martin, we prefer the romance of the shaker.

- **One part vodka**
- **Three parts fresh orange juice**
- **Crushed ice**
- **Slice of orange and maraschino cherry as garnish**

Quarter-fill a cocktail shaker with crushed ice, add the vodka and the orange juice and shake very thoroughly. Strain into a chilled old-fashioned glass, garnish and serve.

WHITE RUSSIAN

You can imagine nostalgic White Russian exiles sipping these after the Bolshevik Revolution and dreaming of the steppes and snows of home. If you leave out the cream and substitute Kahlúa – a rich, brown, coffee-based liqueur from Mexico – for the white Crème de Cacao, you will end up with a Black Russian, which you serve, over ice, in a lowball glass. Try as we might, we have been unable to find a Red Russian to complete the picture.

- **Two parts vodka**
- **One part white Crème de Cacao**
- **One part thick cream**
- **Ice cubes**

Quarter-fill a shaker with ice cubes and add the remaining ingredients. Shake thoroughly and strain into a chilled martini glass. The alternative method, favoured by some U.S. mixologists, is simply to pour the liqueur and the vodka over a couple of ice cubes in an old-fashioned glass and top up with the cream.

Here is a carefully-crafted selection of Martini recipes to try. The greater the proportion of gin to vermouth, the 'drier' the Martini is. Traditionally, you start with the spirit and take it from there.

DRY MARTINI

- **Three parts gin**
- **One part dry vermouth**
- **Ice cubes**
- **A green olive or twist of lemon peel**

Place four ice cubes in a bar glass, followed by the gin and the vermouth. Stir well and strain into a chilled Martini glass. Garnish with the olive on a cocktail stick, or a lemon peel twist.

MEDIUM MARTINI

- **One part gin**
- **One part dry vermouth**
- **One part sweet vermouth**
- **Ice cubes**

You make this as you would a Dry Martini, though it is traditional to double up on the amount of ice and to serve the drink ungarnished.

SWEET MARTINI

- One part gin
- One part sweet vermouth
- Ice cubes
- A maraschino cherry to garnish

You use twice the amount of ice cubes and substitute a maraschino cherry for the olive to garnish the drink. Otherwise, the mixing instructions are the same as for a Dry Martini.

APPLE MARTINI

- Two parts vodka
- One part apple schnapps
- Dash of lime juice
- Ice cubes
- Slice of apple to garnish

Half-fill a cocktail shaker with ice and add the vodka, schnapps and lime juice. Shake vigorously, and then strain into a frosted martini or cocktail glass. Garnish with the apple slice and serve.

WHISKY-BASED COCKTAILS

One whisky is all right; two is too much; three is too few.
Highland saying, cited by Derek Cooper, *A Taste of Scotch*, 1989.

Whisky, say the textbooks, always hails from Scotland, which is why all other varieties should be spelled 'whiskey' to distinguish them from what the Scots rightly regard as their 'water of life'. As far as whisky-based cocktails are concerned, things are a little more complicated by the fact that, as most of them originally came from the U.S.A., the whiskey that they rely on is Bourbon, not Scotch at all. If you want true authenticity, you can obviously use this, but, equally, you should not be put off whisky, for it works just as well in the vast majority of recipes. Other varieties you can try come from Canada and, dare we say it, even from as far afield as Japan. The choice is yours.

right, Manhattan

For reference, though, it is worth noting that Bourbon hails from the American South – from Bourbon County in Kentucky, to be precise. This is where this type of whiskey, which, according to its fans, is more fully-flavoured than Scotch, originated. Bourbon and Bourbon-type whiskeys are also produced in Illinois, Indiana, Ohio, Pennsylvania, Tennessee, Missouri and Virginia. True Bourbon is also what is termed a 'straight' whisky – that is, it has not been blended with any other whiskey – just as, in Scotland, there are also malts and blends. As far as the former are concerned, any Scot worth his or her salt would probably throw a fit if you were to offer them a malt-based cocktail – the way to drink malt is neat, or with a little plain water. Also, never, ever offer a Scot a 'malt on the rocks', if you want to live to tell the tale.

OLD-FASHIONED

An old-fashioned Scot would probably look down his or her nose at any drink which involved 'adulterating' the 'water of life', but where would life be without experimentation?

Try this straightforward traditional recipe for yourself – and we're prepared to bet you'll be won over by its flavoursome bittersweet taste. Some bartenders advocate the use of Canadian Rye, but, as with all things whisky-wise, this is a matter of personal taste.

- A cube of sugar
- Two dashes of Angostura bitters
- Two measures of whisky
- Ice cubes
- A twist of lemon peel

Put the cube of sugar into an old-fashioned or lowball glass, add the bitters and a teaspoon of water and muddle well until the sugar is totally dissolved. Then add the whisky, two ice cubes and and stir well. If you can get any juice out of the lemon peel twist, squeeze this into the drink before you add the twist. Serve with a swizzle stick.

ROB ROY

Back across the Atlantic, whisky comes into its own. Named after the Scots hero immortalized by Sir Walter Scott in his 1817 novel, this tasty cocktail is truly a drink fit for heroes. One word of warning – tasty though these tipples are, do not over-indulge. Otherwise, you might be saying, along with Scott, 'But with the morning cool repentance came'.

- Two parts whisky
- One part sweet vermouth
- Ice cubes
- A twist of orange to garnish

Place six to seven ice cubes in a bar glass, add the whisky and the sweet vermouth and stir well. Strain into a lowball glass, garnish with the orange twist and serve.

THREE RIVERS

This drink originated in Canada, hence the fact that it is often referred to by its French name, 'Trois Rivières'. Perfectionists advocate the use of Canadian Rye for this drink, but do not worry if this is unavailable. The cocktail tastes just as good provided that a good whisky is used.

- **Two parts whisky**
- **One part Dubonnet**
- **One part Triple Sec**
- **Ice cubes**

Quarter-fill a cocktail shaker with ice cubes – five or six should be enough – and then add the whisky and the other two ingredients. Shake well and strain, over more ice cubes, into a lowball glass. Ideally this should be chilled, but this is not essential. Serve ungarnished.

SLOW, COMFORTABLE SCREW

Back again to the American South for this classic, which is a whisky variant on the vodka-based Screwdriver. For once, it is important which whisky you use; only Southern Comfort, the delicious orange-and-peach-flavoured tipple produced in what, way back in the days of Civil War, used to be the Confederate states, gives this cocktail the right flavour.

- **One part Southern Comfort**
- **Six parts fresh orange juice**
- **Ice cubes**
- **Banana slices to garnish**

Half-fill a cocktail shaker with ice cubes. Add the Southern Comfort, followed by the orange juice, and shake thoroughly. Strain the drink into a lowball glass over more ice, garnish with the banana slices and serve.

WHISKY HIGHBALL

Here's a classic from the gallery of all-time highball favourites, all of which are simple and quick to make, as well as being equally delicious to drink. Basically any liquor can be used in combination with ice, soda or other types of mixer.

The American version here uses ginger ale, but soda water would be equally suitable.

- Two parts whisky
- Ginger ale or soda water
- Ice cubes
- A twist of lemon peel

Pour the whisky into a highball glass over the ice cubes – four or five should be enough – and top up with ginger ale or soda water, depending on your taste. Add the twist of lemon peel, stir and serve.

MANHATTAN

The original version of this drink, so it is said, was invented in Maryland in 1846, when it was administered to revive a wounded duellist. This version used sugar syrup, rather than the sweet vermouth featured in the present-day drink, the change coming about in the 1890s, when a Manhattan bartender had the idea for the substitution. That is how the drink got its present name. There are two versions of this cocktail: for a Dry Manhattan, substitute dry vermouth for the sweet variety and an olive for the cocktail cherry.

- **Three parts whisky**
- **One part sweet vermouth**
- **Ice cubes**

Place five or six ice cubes in a bar glass, add the whisky and the sweet vermouth, stir well and strain. Pour the drink into a chilled martini or cocktail glass.

MINT JULEP

In the Deep South of the U.S.A – where this refreshing brew originated back in the days of slavery – time, labour and cost were the last things wealthy cotton barons had to take into account; and the Mint Julep became a favourite plantation tipple. However, it is a time-consuming cocktail to prepare, as well as being quite an expensive one, so it's really only worth the effort if you are having a few friends around to enjoy a snorter or two. Also, there is really no substitute for Bourbon – preferably from Kentucky – in this particular instance, though some people say that you can use whisky. Others advocate the use of dry gin, rum and, most notably, brandy.

• A tankard of Bourbon
• A teaspoon of light rum
• Two tablespoons of water

- A teaspoon of sugar
- Crushed ice and ice cubes
- A large bunch of fresh mint sprigs

Put plenty of the mint sprigs into a mixing glass, dust with the sugar and add the water, which should be just enough to dissolve the sugar. Muddle the mint gently – the aim is to crush the leaves just enough to release their flavour – then add around a cupful of crushed ice, followed by the Bourbon and finishing off with the rum. Stir gently, then strain into a serving jug, which should be refrigerated until required. When serving individual drinks, which you do in Collins or lowball glasses, put a couple of ice cubes in each glass before adding the liquor, and garnish with a sprig or two of mint, the leaves of which have been dipped in powdered sugar. It is traditional to sip the Julep daintily through a straw, rather than quaffing it.

NEW ORLEANS

There's no dispute about where this tangy cocktail originated, right down in the heart of what Southerners still affectionately call Dixieland. Maybe its creation owed something to the festival spirit of Mardi Gras and the seductive, throbbing rhythms of traditional New Orleans jazz.

Quarter-fill a bar glass with ice cubes, shake the bitters over the ice and then add the other ingredients. Stir thoroughly and then strain into a lowball glass, in which you have already placed an ice cube. Garnish with the mint and serve.

WALDORF

Another favourite hang-out for the New York cocktail crowd was the bar of the Waldorf Hotel. The original recipe is based on Bourbon, but there is nothing to stop you trying other whiskies to create variants of your own.

- Two parts Bourbon
- One part Pernod
- One part sweet vermouth
- A dash of Angostura bitters
- Crushed ice

Put some crushed ice into a bar glass – a scoop of ice should be enough in this instance – and add the Bourbon, Pernod, sweet vermouth and the dash of bitters. Stir well. Strain into a chilled cocktail glass and serve just as it is, without any garnish.

BRAINSTORM

The next time you're trapped in a meeting room, surrounded by flip charts and people desperately trying to think up new ideas, why not suggest that one or two of these might serve to lubricate the creative process nicely?

- **One part whisky**
- **One part dry vermouth**
- **One part Benedictine**
- **Crushed ice**
- **A slice of orange**

Put some crushed ice into a bar glass – two to three tablespoons will be enough – followed by the whisky, dry vermouth and Benedictine, which you pour over the ice. Stir thoroughly and then strain into a chilled cocktail or tulip glass. Garnish with the orange slice and serve.

RUSTY NAIL

It's back to Scotland again for this traditional favourite. The whisky, without question, must be Scotch, while the Drambuie, itself Scotch whisky-based, brings its own individual flavour of whisky, heather and honey to the party.

• Two parts Scotch
• One part Drambuie
• Ice cubes
• A twist of lemon to garnish

Fill an old-fashioned glass with ice cubes, add the whisky, and then float the Drambuie over the top. Serve garnished with the lemon twist.

REBEL CHARGE

It's back to the American South for this final offering. In the days of Civil War, the rebel yell of the Confederate troops as they charged was often enough to put the fear of God into their Union opponents. What history doesn't record is whether or not Dixie spirit was fuelled by this potent cocktail.

- Two parts Bourbon
- One part Triple Sec
- A tablespoon of orange juice
- A tablespoon of lemon juice
- Half an egg white
- Ice cubes
- A slice of orange

Quarter-fill a cocktail shaker with ice cubes, followed by the Bourbon and the other ingredients, ending up with the half egg white. Shake well and strain into an old-fashioned glass over more ice cubes. Add the slice of orange and then serve.

CHAMPAGNE-BASED COCKTAILS

After all, what is your host's purpose in having a party? Surely not for you to enjoy yourself; if that were their sole purpose, they'd have simply sent champagne and women over to your place by taxi.

P. J. O'Rourke, American writer and humourist, attrib.

First of all, praise is due to Dom Perignon, the 17th-century French Dominican monk whose discovery of what is termed méthode champenoise made champagne possible. Essentially, what this involves is allowing the still wine on which the champagne is based to ferment a second time in the bottle, a process which produces the drink's characteristic bubbles.

Of course, this base wine – what Frenchmen call the cuvée – has to be as near perfect as possible in itself: by law, only Chardonnay or Pinot Noir grape varieties can be used for it, either individually or in a blend. If

right, Buck's Fizz

the label on the bottle says Blanc de Noir, this means
that only Pinot Noir grapes have been employed,
while, if the term Blanc de Blanc is used, the wine is a
hundred per cent Chardonnay. Each wine is also
labelled according to level of sweetness: brut is the
driest, sec is slightly sweeter and demi-sec is fairly
sweet. As far as vintages and non-vintages go, every
champagne house produces its own standard non-
vintage bottling, consisting of a blend of several
vintages. It's only in exceptional years that a vintage-
dated wine, made solely from grapes from that
particular harvest, is produced.

At its best, champagne is one of the most versatile of
beverages, its tasty, lemony, elegant flavours going
well with practically every variety of liqueur and fruit
juice. Mixologists have taken advantage of this to
create a host of excitingly effervescent cocktails, all of
which do full justice to the unique regal qualities of
this 'king of wines'.

CHAMPAGNE COCKTAIL

Some people swear that the brandy makes all the difference to this classic, others leave it out. The choice is yours. As a rough guide, one bottle of champagne should make around six cocktails. If you are feeling extravagant, a pitcher of cocktails – what mixologists call Champagne Cup – is just the thing to produce for a group gathering, in which case you can save on champagne by diluting the mixture slightly with soda water, making up for this by adding two parts of brandy and one part of Triple Sec to the mix. Garnish the jug with some seasonal fruits and slices of cucumber and top with a small bunch of mint. You will need to add some more sugar – use powdered sugar, rather than cubes in this instance – but exactly how much is down to personal taste.

- A sugar cube
- Two dashes of Angostura bitters
- Chilled champagne (the drier the better)
- A teaspoon of brandy

Place the sugar cube in a champagne flute (you could use a champagne saucer, but this will mean less to drink). If you can chill the flute in advance, so much the better. Splash the Angostura bitters onto the sugar cube, add the champagne and float the brandy on top.

BELLINI

Created by Giuseppe Cipriani, the founding father of Harry's Bar in Venice, and named in honour of the great Venetian artist, this became a favourite tipple of a host of celebrity visitors, including the writer Ernest Hemingway and the actor and playwright Noel Coward. Many misguided attempts have been made to improve on the original recipe – substituting canned peach juice for the freshly-squeezed juice the recipe calls for and adding dashes of lemon and Grenadine – but, in this instance, such tinkering simply doesn't pay. For the peach juice, purée a peeled peach in a blender.

- One part fresh peach juice
- Three parts dry champagne
- A peach ball to garnish

left, Ernest Hemingway

Pour the fruit juice into a chilled champagne flute until the flute is about a quarter full. Top up the glass with the champagne, garnish with the peach ball and serve. Simple!

KIR ROYALE

The original, more plebeian Kir was invented by a French priest of that name, who decided to experiment by adding some Crème de Cassis to a glass of white Burgundy. This regal version is altogether more grand.

• Dry champagne
• A dash or two of chilled Crème de Cassis
• A twist of lemon rind as garnish

Three-quarters fill a champagne flute with the champagne – this should be well-chilled – splash in the cassis, decorate the glass with the twist of lemon and serve.

BUCK'S FIZZ

Named after its founder-president, Captain Buckmaster, Buck's is one of the most exclusive of London's gentlemen's clubs. This is the club version of the French Champagne à l'Orange, devised by the club's head barman back in the 20s. Because one glass of this fine fizz simply doesn't go far enough, use the proportions guide to make a full jug if you are entertaining.

- One part fresh orange juice
- Two parts dry champagne (a good sparkling wine could be substituted in this instance)
- Orange wedge and cocktail cherry

Pour the orange juice into a chilled champagne flute and top up with the chilled dry champagne. Stir very gently, garnish with the orange twist and serve at once.

BLACK VELVET

Across the Atlantic, there appears to be the impression that any old 'chilled stout' will suffice for this, but, as any Irishman will tell you, only Guinness – preferably from the original Dublin brewery on the Liffey – will really do. The drink is said to have been created in Brooks' Club, another London gentlemen's watering-hole, in 1861, as the nation mourned the death of Prince Albert, Queen Victoria's Prince Consort. Traditionally, this delicious brew was served in a silver beer tankard, but, in these less extravagant times, an ordinary glass will do. It is not a drink to linger over or nurse – the secret is to quaff it as soon as it has been made.

- **One part Guinness stout (canned or bottled)**
- **One part dry champagne**

Half-fill the glass with the stout, tilt it to the side and pour in the champagne gently, trying to create as little foam as possible. The champagne should be well-chilled and, if the Guinness itself has a head on it, it is best to let this settle first. That's all.

ROSSINI

If you like your Bellinis, you'll adore their near relation, the Rossini, though, on this occasion, we don't know whether the name honours the great 19th-century Italian opera composer or not. As Rossini himself was a great gourmet – witness the Tournedos Rossini that we know were named after him – it would be nice to think that he would have appreciated this refreshing drink as well.

• One part puréed strawberries
• Three parts dry champagne
• A fresh strawberry as garnish

Purée the strawberries in an electric blender, pour into a chilled champagne flute and top up with the champagne. Stir very gently – if you stir too vigorously, you will dissipate the bubbles – and float the fresh strawberry on the top. Serve immediately.

DEATH IN THE AFTERNOON

Reputed to have been one of Ernest Hemingway's favourite drinks from the time when, as a young writer, he took up residence in Paris, this powerful concoction must have been responsible in its time for many 'lost' afternoons, if not for any actual deaths. You need to be careful when you make it as otherwise you run the risk of losing the delicate champagne bubbles.

- **Dry champagne**
- **One part Pernod**
- **Ice cubes**

Place a couple of ice cubes in the bottom of a champagne flute and pour the Pernod over them. Gently top up the glass with the champagne – the colder this is the better – and then stir the completed drink equally gently. The aim is to mix the Pernod with the champagne without losing the latter's sparkle. Serve just as it is, ungarnished.

TYPHOON

You'll have noticed that, in all the other cocktail recipes given in this chapter, we advise very gentle stirring at best, and avoid the use of a cocktail shaker for obvious reasons. This heady American brew is the exception. So batten down the hatches and prepare to ride out the storm!

- **One part gin**
- **Half a part Anisette**
- **One part lime juice**
- **Dry champagne**
- **Ice cubes**

Quarter-fill a cocktail shaker with ice cubes, add all the other ingredients with the exception of the champagne, and shake well. Strain into a Collins glass, over ice, and then top up the glass with the champagne. As always, the champagne should be well-chilled in advance.

ARISE MY LOVE

Simplicity itself to make, this colour-rich cocktail is the perfect prelude to a romantic evening à deux.

- **A tablespoon of green Crème de Menthe**
- **Dry champagne**

Pour the liqueur into the bottom of a chilled champagne flute, and fill the flute with chilled champagne. Stir very gently until the two are mixed, taking care not to disperse the champagne bubbles, and serve.

AMBROSIA

The ancient Greeks believed that ambrosia, a kind of honey mead, was the drink of the Gods, who lived high up on the summit of Mount Olympus. This modern-day version from across the Atlantic slips down equally nicely.

The original recipe calls for applejack (American Apple Brandy), but we have substituted Calvados for it. Again, you will need to use a cocktail shaker for the first part of the recipe

- One part Calvados
- One part brandy
- A dash of Triple Sec
- Juice of a lemon
- Dry champagne
- Ice cubes

Quarter-fill a cocktail shaker with ice cubes, add all the ingredients except for the champagne and shake thoroughly. Pour the contents into a highball glass over more ice – there is no need to strain them – and top up with chilled champagne.

SHOOTERS, SLAMMERS AND TEQUILA

A taste for drink, combined with gout,
Had doubled him up for ever.
Of that there is no manner of doubt –
No probable, possible shadow of doubt –
No possible doubt whatever.
W. S. Gilbert, *The Gondoliers*

Whether the fate of the late ruler of Gilbert and Sullivan's fantasy island of Barataria would have been speedier had he been a hardened tippler of shooters, slammers and tequila is not up to us to say for certain, but it would seem highly probable.

Shooters are a relative newcomer to the cocktail scene, invented, so they say, in Canada by an ingenious group of barmen as a prescription against the cold – and boredom – of those long Canadian winter nights. The idea is to float ingredients of different colours on top of one another, the key to success being to start with the heaviest ingredient first and end up with the lightest. The shooter and

right, Margarita

slammer tradition calls for the results to be knocked back in a single gulp, but it's up to you whether you can live with the possible consequences. Tequila, too, is a relatively new entrant onto the global cocktail scene, though the Margarita is now undoubtedly a cocktail classic.

POUSSÉ CAFÉ

This sweet, striped wonder is the original from which the whole idea of shooters sprang. The trick to making one successfully is to build the drink carefully, starting with the heaviest constituent and pouring each of the subsequent ones over the back

of a barspoon, so each colour remains separate until you tilt the glass as you drink from it.

- **One part Grenadine**
- **One part Yellow Chartreuse**
- **One part Crème de Cassis**
- **One part white Crème de Menthe**
- **One part Green Chartreuse**
- **One part brandy**

Following the order given, pour each ingredient in turn into a poussé café glass, so that each ingredient floats on the preceding one. This demands a steady hand and the back of a barspoon.

TEQUILA SLAMMER

In the U.S., this cocktail is often referred to as a Tequila Straight, for patently obvious reasons. If you visit Mexico, this is how tequila is often drunk in clubs and bars, though, over here, the tradition is now to follow the slammer with a beer chaser.

- A measure of tequila
- A pinch of salt
- A quarter of a lemon

This is a two-handed drink, for reasons that will become obvious. Put the salt between the thumb and the index finger on the back of your left hand. Holding a jigger-full of tequila in the same hand and the quarter of lemon in your right, lick the salt off your hand, down the tequila, and then suck at the lemon.

MARGARITA

Salt plays its part in the composition of a Margarita as well, though here its role is limited to serving as a frosting around the rim of the glass in which the cocktail is served.

- **Three parts tequila**
- **One part Triple Sec**
- **One part lime or lemon juice**
- **Ice cubes**
- **Salt**

Rub the rim of a cocktail glass with the lemon or lime rind, frost the rim with the salt and set aside. Quarter-fill a cocktail shaker with ice cubes, add the tequila, Triple Sec and the fruit juice and shake well. Strain, over an extra couple of ice cubes, into the cocktail glass, taking care not to disturb the salt frosting. Serve immediately.

TAMARIND MARGARITA

- ½ cup sugar
- 4½ cups water
- Wet tamarind (cut into cubes)
- One part Cointreau
- One part tequila
- One part fresh lime juice
- Ice cubes
- Lime slices to garnish

Mix the sugar and water together in a small saucepan over a medium heat. Stir until the sugar dissolves and then leave to cool. In another saucepan combine the remaining water and the tamarind cubes until simmering. Remove and reserve for about two hours until the tamarind is really soft. Transfer to a sieve with a bowl underneath and reserve the soaking liquid. Press the pulp through the sieve, using the back of a wooden spoon, into a different bowl. There should be around 1 cup of pulp; if not top up with the reserved soaking liquid. Blend half the ice cubes, lime juice, tamarind pulp and sugar syrup until slushy and pour equal amounts into martini glasses. Repeat with the Cointreau, tequila and remaining ice, lime juice, tamarind pulp and sugar syrup, and top up. Garnish with lime slices and serve.

TEQUILA SUNRISE

Another cocktail classic from Mexico, fiery tequila's tropical homeland. The spirit is, in fact, distilled cactus juice, but don't let that put you off. The name itself comes from an early Spanish settlement, where the first distillery to make the heady brew was established many, many years ago.

- **One part tequila**
- **Three parts fresh orange juice**
- **Two dashes of Grenadine**
- **Ice cubes**
- **Lemon slice and cocktail cherry to garnish**

Half-fill a cocktail shaker with ice cubes, followed by the tequila and orange juice. Shake well and strain the mixture, over ice, into a highball glass. Add the Grenadine – pour this slowly, as the idea is to allow the colour to sink slowly through the drink. Allow the Grenadine to settle, garnish and serve.

STEADY THE BUFFS!
NON-ALCOHOLIC
COCKTAILS

Abstinence is as easy to me as temperance would be difficult.
Dr. Samuel Johnson

While we tend to share the viewpoint of the good doctor, there is no getting around the fact that, these days, any mixologist worth his or her salt should have a few good non-alcoholic cocktail recipes tucked up their sleeves, ready for use when the occasion calls. One such instance would be if you – or any of your guests – are driving, especially with the drink-driving laws rightly getting ever tighter and tighter. There may be other reasons for someone abstaining from alcohol as well; if you are taking antihistamines for hay fever, for instance, many doctors will advise you not to drink, as the combination may make you sleepy.

right, Ginger Apple Punch

Basically, though, it's now considered good bar manners to have a non-alcoholic option on offer, so the trick is to find – or devise – some that will provide pleasure to any of your guests who want to take this route. They'll be doubly grateful if you do this – no-one wants to stand around holding a glass of sad-looking mineral water and looking like a complete party-pooper. As with any drink, the secret in this instance is to find a good balance between the extremes of sweetness and acidity for the brews you serve to be a runaway success.

VIRGIN MARY

This is a tried-and-tested standby that is always certain to please. All you do is to take the 'bloody' out of the Bloody Mary by removing the alcohol.

• Tomato juice
• Half a part of lemon juice
• A dash of Tabasco sauce per glass
• A dash of Worcestershire sauce per glass
• Celery salt
• Freshly ground white pepper
• Ice cubes
• A stick of celery as garnish

Half-fill a cocktail shaker with ice cubes, add the tomato and lemon juice, followed by the other seasonings, and shake well. Strain into a highball glass, garnishing the drink with the celery stick, which you can use as a stirrer.

left, Virgin Mary

SHIRLEY TEMPLE

Child-star Shirley Temple, the Hollywood *Wunderkind* of her day, later went into politics, eventually becoming an American Congresswoman. This cocktail is simple and straightforward, though maybe, like Shirley Temple herself, a little too sweet for some tastes.

- Ginger ale
- A dash of Grenadine
- Ice cubes
- An orange slice and marashino cherry to garnish

Put plenty of ice cubes into a highball glass, dash the Grenadine over them and then add the ginger ale. Stir slightly before garnishing the drink with the orange slice and maraschino cherry and serve.

right, Shirley Temple

ST CLEMENTS

It's the old nursery song that gave this cooling cocktail its name – and, with this sure-and-safe recipe, there's absolutely no risk at all of an unwelcome chopper coming to 'chop off your head'.

- **Orange juice**
- **Bitter lemon**
- **Crushed ice**
- **Lemon and orange slices as garnish**

Fill a cocktail shaker a quarter full of crushed ice, followed by orange juice. Shake well and pour the results into a Collins or highball glass. Add the bitter lemon – you need as much bitter lemon as you have of the orange juice – and stir. Garnish with the lemon and orange slices and serve, with a straw.

GINGER APPLE PUNCH

Most fruit punches combine well to produce interestingly-flavoured, thirst-quenching drinks. This example is no exception to the rule.

• **Apple juice**
• **One sliced dessert apple**
• **Ice cubes**
• **Dry ginger ale**

Place a carton of apple juice along with the apple slices in a jug or punch bowl. Add plenty of ice and top up with ginger ale – you need to use as much ginger ale as apple juice.

HANGOVER CURES

There is no hangover on earth like the single malt hangover. It roars in the ears, burns in the stomach and sizzles in the brain like a short circuit. Death is the easy way out.
Ian Bell, *The Observer*, 1991

Amongst other things, P.G. Wodehouse's ever-resourceful Jeeves was renowned for his ability to mix the perfect hangover cure. In fact, it was this that persuaded man-about-town Bertie Wooster to hire him on the spot that morning when the two met for the first time in *Jeeves Takes Charge*:

'If you would drink this, sir,' he said, with a kind of bedside manner, rather like the royal doctor shooting the bracer into the sick prince. 'It is a little preparation of my own invention. It is the Worcestershire Sauce that gives it its colour. The raw egg makes it nutritious. The red pepper gives it its bite. Gentlemen have told me that they have found it extremely invigorating after a late night out.'

right, Prairie Oyster

I would have clutched at anything that looked like a lifeline that morning. I swallowed the stuff. For a moment I felt as if somebody had touched off a bomb inside the old bean and was strolling down my throat with a lighted torch, and then everything seemed suddenly to get all right. The sun shone in through the window; birds twittered in the tree-tops; and, generally speaking, hope dawned once more.

'You're engaged!' I said, as soon as I could say anything. I perceived clearly that this cove was one of the world's workers, the sort that no home should be without.

While we would not claim to be as omniscient as the all-wise, all-seeing Jeeves – not for nothing has his name passed into the language as the epitome of the perfect gentleman's gentleman – the recipes that follow here are all well-tried tonics and pick-me-ups that have stood the test of circumstance.

PRAIRIE OYSTER

In common with most hangover cures, the tradition here is to down the drink in one!

- **One part brandy**
- **Two teaspoons of cider vinegar**
- **A dessert-spoonful of Worcestershire sauce**
- **A teaspoon of tomato ketchup**
- **A dash of Angostura bitters**
- **The yolk of a fresh egg**
- **Cayenne pepper**
- **Ice cubes**

Quarter-fill a cocktail shaker with ice cubes, add the brandy, vinegar, Worcestershire Sauce, ketchup and bitters and shake thoroughly. Strain into a lowball glass over two more ice cubes – the liquid should nearly reach the top of the glass. Float the egg yolk on top of the drink, taking care not to break it, and sprinkle it lightly with a pinch or two of cayenne pepper.

BULLSHOT

Some people claim that this is one of the quickest and most effective hangover cures around, but the bouillon is an acquired taste. You either like it, or you don't. Certainly, it's not a drink for a vegetarian.

- **One part vodka**
- **Two parts clear chilled beef bouillon**
- **A dash of Worcestershire sauce**
- **A dash of lemon juice**
- **A pinch of salt**
- **A pinch of pepper**
- **Ice cubes**

Put some ice cubes into a cocktail shaker – five or six will be enough – and add the vodka and the beef bouillon, followed by the Worcestershire sauce and the dash of lemon juice. Season with the salt and pepper and then shake thoroughly. Strain, over ice, into a lowball glass.

SUNBURST

Finally, in case you are starting to think that we are dedicated proponents of the 'hair of the dog' school of thought when it comes to the dreaded morning-after, here is a non-alcoholic pick-me-up that is positively awash with healthy vitamins. The only thing that you will have to grit your teeth over is the noise of the blender. Sorry!

- A green apple, cored and chopped
- Three peeled and chopped carrots
- A peeled and stoned mango
- Seven parts chilled, fresh orange juice
- Hulled strawberries
- Slice of orange as garnish

Place the apple, carrots and mango in the blender bowl and process to a pulp. Add the orange juice and the strawberries and process again. Strain through a sieve, pressing out all the juice from the residue of pulp the sieve will retain with the back of a wooden spoon. Pour into a highball glass, over ice, garnish with the slice of orange and serve.

INDEX

Acapulco 53
Algonquin 142–3
Algonquin Hotel, New York 142
Ambrosia 163
Apple Martini 127
Arise My Love 161
Au Revoir 64

B&B 61
B&B Collins 61
Bacardi 45
Banana Daiquiri 49
'bathtub' gin 11
Bee's Knees 46
Bellini 153–4
Between the Sheets 50
Black Marble 116–17
Black Velvet 156
Blacksmith Cocktail 76
blenders 29
Bloody Maria 112
Bloody Mary 111–12
Bloody Muddle 112
Blue Devil 94–5
Blue Lagoon 113
Bombay Cocktail 77

bootlegging 10–11
Bosom Caresser 75
Bourbon 18, 128–30
Brainstorm 145
brandy 20, 58–60
Brandy Alexander 62–3
Brandy Classic 67
Brandy Sour 69
Brass Monkey 40–1
Bronx 96
Brooks' Club, London 156
Buck's Fizz 155
Bullshot 185–6

Cardinale 89
Caruso 93
Caruso, Enrico 92, 93
champagne 21, 148–50
Champagne Cocktail 151–3
Charleston 63
Charlie Chaplin 64
Cipriani, Giuseppe 153
Classic Daiquiri 48
Classic Martini 124
cock-ale 8
cocktail parties 12
cocktail shakers 29

cocktails:
  history 8–12
  ingredients 17–26
  preparing 14–15
  secrets of 13–15
Coffee 73
'coffin varnish' 11
Collins, John 97
Connolly, Charley 105
Conrad, Joseph 100
coquetel 8
Corkscrew 44
Cosmopolitan 115
Coward, Noel 12, 153
Cuba Libre 43

Daiquiri 48–9
Death in the Afternoon 159
Dry Martini 12, 13–14, 126

Edinburgh 143–4
Eggnog 66

fixings 25–6
Flanagan, Betty 8
Fleming, Ian 112, 120, 124
French '75 102–3

Gibson 105
Gibson, Charles Dana 105
Gilbert, W.S. 164

Gimlet 103–4
gin 18, 78–80
Gin Alexander 97
Gin Fizz 88
Gin and Sin 85
Gin Sling 84
Gin Smash 87
Gin Sour 86
Ginger Apple Punch 181
glasses:
  chilling 27
  frosting 33
  for mixing 30
  types 31–3
Green Devil 104–5
Grenadier 72

hangover cures 182–7
Harry's Bars 12, 91, 102,
  153
Harvard 74
Harvey Wallbanger 118
Hemingway, Ernest 12, 52,
  152, 153, 159
Hoover, Herbert 10
Horse's Neck 71

ice 27
ingredients 17–26

Johnson, Dr Samuel 58, 174

Kir Royale 154
kit 28–31
Knickerbocker Hotel, New York 124

Limmers Hotel, London 97
liqueurs 21–4
Long Island Tea 57
Long Vodka 121

Mai Tai 56
Maiden's Prayer 102
Manhattan 138
Margarita 170
Martin, John 114, 122
Mary Pickford 54, 55
Maugham, Somerset 100, 101
measures 29–30
Medium Martini 126
Mint Julep 140–1
mixers 24–5
mixologists 11
Mojito 52
'moonshine' 11
Moscow Mule 114

Negroni 90–1
New Orleans 141–2
non-alcoholic cocktails 174–6

Old-Fashioned 131–2
Orange Cosmo 116

Parker, Dorothy 142–3
Petiot, Fernand 111
Pickford, Mary 54, 55, 111
Piña Colada 44
Pink Gin 107
Pink Lady 81
Planter's Punch 39–40
Poussé Café 166–7
Powell, Anthony 90
Prairie Oyster 184–5
Prohibition 10

Raffles Hotel, Singapore 12, 100
Rebel Charge 147
Ritz, London 11
Rob Roy 132
Rockefeller, John D. 124
Rossini 158
'rotgut' 11
rum 18, 36–8
Rusty Nail 146

St Clements 180
St Regis Hotel, New York 11
Salty Dog 117
Savoy, London 11
Screwdriver 122